The United States Army War College

The United States Army War College educates and develops leaders for service at the strategic level while advancing knowledge in the global application of Landpower.

The purpose of the United States Army War College is to produce graduates who are skilled critical thinkers and complex problem solvers. Concurrently, it is our duty to the U.S. Army to also act as a "think factory" for commanders and civilian leaders at the strategic level worldwide and routinely engage in discourse and debate concerning the role of ground forces in achieving national security objectives.

The Strategic Studies Institute publishes national security and strategic research and analysis to influence policy debate and bridge the gap between military and academia.

The Center for Strategic Leadership contributes to the education of world class senior leaders, develops expert knowledge, and provides solutions to strategic Army issues affecting the national security community.

The Peacekeeping and Stability Operations Institute provides subject matter expertise, technical review, and writing expertise to agencies that develop stability operations concepts and doctrines.

The School of Strategic Landpower develops strategic leaders by providing a strong foundation of wisdom grounded in mastery of the profession of arms, and by serving as a crucible for educating future leaders in the analysis, evaluation, and refinement of professional expertise in war, strategy, operations, national security, resource management, and responsible command.

The U.S. Army Heritage and Education Center acquires, conserves, and exhibits historical materials for use to support the U.S. Army, educate an international audience, and honor soldiers—past and present.

STRATEGIC STUDIES INSTITUTE

The Strategic Studies Institute (SSI) is part of the U.S. Army War College and is the strategic-level study agent for issues related to national security and military strategy with emphasis on geostrategic analysis.

The mission of SSI is to use independent analysis to conduct strategic studies that develop policy recommendations on:

- Strategy, planning, and policy for joint and combined employment of military forces;
- Regional strategic appraisals;
- The nature of land warfare;
- Matters affecting the Army's future;
- The concepts, philosophy, and theory of strategy; and,
- Other issues of importance to the leadership of the Army.

Studies produced by civilian and military analysts concern topics having strategic implications for the Army, the Department of Defense, and the larger national security community.

In addition to its studies, SSI publishes special reports on topics of special or immediate interest. These include edited proceedings of conferences and topically oriented roundtables, expanded trip reports, and quick-reaction responses to senior Army leaders.

The Institute provides a valuable analytical capability within the Army to address strategic and other issues in support of Army participation in national security policy formulation.

Strategic Studies Institute
and
U.S. Army War College Press

UNLOCKING INDIA'S STRATEGIC POTENTIAL IN CENTRAL ASIA

Roman Muzalevsky

October 2015

*****

Comments pertaining to this report are invited and should be forwarded to: Director, Strategic Studies Institute and U.S. Army War College Press, U.S. Army War College, 47 Ashburn Drive, Carlisle, PA 17013-5010.

*****

This manuscript was funded by the U.S. Army War College External Research Associates Program. Information on this program is available on our website, *www.StrategicStudiesInstitute.army.mil*, at the Opportunities tab.

*****

*****

The Strategic Studies Institute and U.S. Army War College Press publishes a monthly email newsletter to update the national security community on the research of our analysts, recent and forthcoming publications, and upcoming conferences sponsored by the Institute. Each newsletter also provides a strategic commentary by one of our research analysts. If you are interested in receiving this newsletter, please subscribe on the SSI website at *www.StrategicStudiesInstitute.army.mil/newsletter.*

ISBN 1-58487-705-7

CONTENTS

FOREWORD

As a region with abundant resources and rapidly growing transit potential surrounded by nuclear-armed powers, Central Asia is increasingly drawing the attention of global players. Russia is actively seeking to rebuild its economic influence via the newly created Eurasian Economic Union. China is expanding its reach through a recently launched Silk Road Economic Belt. Other actors are jockeying for their share of the region's pie, as well. But the United States and India are enjoying only very limited presence in what is increasingly becoming a critical part of the world.

In this comprehensive and insightful account, Mr. Roman Muzalevsky, an author of a book and several monographs on global trends, great power politics, grand strategy, and connectivity issues, explains why India lags behind other actors in the region and what needs to be done to unlock its potential as a rising great power and shore up its strategic presence in Central Asia. The region, he argues, is of growing importance for India's expansion as an emerging continental power, and failure to enhance its footprint risks delaying India's global rise and undermining the U.S. global agenda of upholding the global order amid accelerating power shifts.

According to the author, a number of select elements of India's strategic culture and geopolitical constraints have prevented Delhi from pursuing a more active and effective regional policy. India's legacy of nonalignment, lost orientation, and inward focus following the collapse of the Soviet Union, among other factors, are, in part, responsible for the country's lack of clear direction, absence of a widely appealing model, and somewhat passive foreign policy that still draws heavily on the outdated framework of India's nonalignment legacy as its baseline.

As if it were not enough, Mr. Muzalevsky contends that India also confronts major geopolitical constraints such as a disconnected Central-South Asian region, instability and volatility in Afghanistan and Pakistan, and tensions and border disputes with Beijing and Islamabad. China's head start in undertaking economic reforms and its perceived strategy of encirclement of India have added further strains on India's ambitions to enhance its strategic profile in Central Asia. The author approaches each of these issues with a critical eye and through the prism of India's relations with individual Central Asian republics and great powers, highlighting deficiencies of India's overall approach to the region and challenges and opportunities of its "Connect Central Asia" policy. He argues that India needs not only to compete but also cooperate with its perceived rivals in the broader region, especially China and Pakistan. Most importantly, however, Mr. Muzalevsky calls for an explicit partnership between the United States and India in the region, pointing to their overlapping agendas as part of the U.S. New Silk Road Strategy and India's "Connect Central Asia" policy, which, in many ways, confront similar constraints.

Mr. Muzalevsky treats the prospects of such partnership with caution, pointing to India's disinterest in power balancing schemes and proclivity for pursuing a strictly independent course. He also highlights the potential of this partnership to undermine U.S.-Russian and U.S.-Chinese relations and lead to escalation of external rivalries in a region that is barely able to cope with internal ones. But the author also provides compelling reasons for why such partnership is a must if Delhi and Washington want to advance their interests in and out of Central-South Asia. Neither of these powers enjoys a substantial presence in the region, and each is poorly positioned to take advantage of regional trends and opportunities to shape Russia's, China's, and Iran's advances. Both, however, are English-speaking democracies, concerned about the future of Central-South Asia, given regional threats, Russia's resurgence, and China's expansion. Both are also seeking to reverse decades of separation between Central and South Asia as part of their strategies.

The parties are already aligning their positions on select issues and see a growing cooperation on strategic matters, including nuclear issues and privatization and investment into the Indian economy. The author makes an important caveat when discussing the prospects of such partnership in Central Asia, stating that the United States and India can now start pursuing a number of objectives, either in concert or separately, to unlock each other's strategic potential in the region and beyond. Specifically, they should mitigate Afghanistan-Pakistan security challenges; shape Iran's geopolitical role; foster Sino-Indian cooperation; and exploit grand strategies and regional connectivity initiatives of other players in Central-South Asia and the Greater Middle East. But the author's advice stands: Delhi and Washington should partner to remain relevant in the region.

The Strategic Studies Institute is pleased to offer this insightful policy guide to its audience of international relations professionals and policymakers working on issues ranging from U.S. and Indian policy in

Central-South Asia to U.S.-Indian, Sino-Indian, and Indo-Pakistani relations.

DOUGLAS C. LOVELACE, JR.
Director
Strategic Studies Institute and
U.S. Army War College Press

ABOUT THE AUTHOR

ROMAN MUZALEVSKY works for iJet International Inc., performing research, reporting, and analyzing global, regional and national security affairs, risks, and trends, as well as providing incident response and crises management services. He is also a Contributing Analyst on Eurasian Affairs and Security at Jamestown Foundation. Previously, he worked for CSM Solutions Inc., National Democratic Institute, Department for International Development Public Finance Reform Project, Central Asia-Caucasus Institute, security consultancy Wikistrat, and global strategy advisory consultancy Krull Corp. in the United States and Central Asia. He conducted research on security issues in the post-Soviet space at the Center for Political and Military Analysis at Hudson Institute as part of the George F. Jewett Foundation Fellowship Award for Projects on the Study and Practice of Grand Strategies. Mr. Muzalevsky has authored at least 100 articles on Eurasian and global affairs, geopolitics, security, and strategic trends. He is the author of *Central Asia's Shrinking Connectivity Gap: Implications for U.S. Strategy* (2014); *From Frozen Ties to Strategic Engagement: U.S.-Iranian Relationship in 2030* (2015); and *China's Rise and Reconfiguration of Central Asia's Geopolitics: A Case for U.S. Pivot to Eurasia* (2015). Mr. Muzalevsky holds a diploma in international affairs from the International Ataturk Ala-Too University in Kyrgyzstan and an M.A. in international affairs with concentration in Security and Strategy Studies from Yale University.

SUMMARY

India's impressive economic growth over the last 2 1/2 decades has brought India's role and interests to the forefront of global politics and statecraft. Importantly, it has put India into a comparative perspective with China, another aspiring Asian great power poised to stiffen competition for resources and influence worldwide. Both are resource-hungry and rapidly emerging powers seeking a new place and role in the global and regional orders. Both are also strategic rivals and consider their immediate neighborhood of Central Asia of growing strategic importance to their grand strategies. For now, China has outperformed India in Central Asia on all counts, securing the region as a key resource base and platform for power projection. India launched the "Connect Central Asia" policy in 2012 to shore up its presence, but the policy has not yet secured for it even a remotely comparable stake in the region due to aspects of India's strategic culture and geopolitical constraints. Meanwhile, U.S. strategic presence in the region leaves much to be desired. The United States is withdrawing from Afghanistan without major political or military gains from the conflict that has cost it and its partners a fortune in lives and money. The future of its military infrastructure and relationships with countries in Central-South Asia is a big unknown, with regional partners equating the U.S. military pullout with its waning commitment to support the regional economic and security order. To help unlock their strategic potentials, Delhi and Washington should join forces and cultivate a strategic partnership that makes Central Asia its major pillar. Until then, neither Delhi nor Washington is likely to succeed. Written in May 2015, this monograph examines related issues and proposes regional policy recommendations for India and the United States.

UNLOCKING INDIA'S STRATEGIC POTENTIAL IN CENTRAL ASIA

INTRODUCTION

India's impressive economic growth over the last 2 1/2 decades has brought India's role and interests in the world and Central-South Asia to the forefront of global politics and statecraft. Importantly, it has put India into a comparative perspective with China, another aspiring Asian great power poised to stiffen competition for resources and influence among established and emerging powers alike. Both are resource-hungry and rapidly emerging powers seeking a new place and role in global and regional economic and security orders. Both are also strategic rivals and consider their immediate neighborhood of Central Asia of growing strategic importance to their grand strategies. Just as China has viewed its westward expansion into Central Asia imperative to acquire resources and project power globally, so has India deemed its northward push into the region essential to nurture and expand its potential as an aspiring global power.

For now, China has outperformed India in Central Asia on all counts, securing the region as a key resource base and platform for power projection. In 2012, India launched the "Connect Central Asia" policy to shore up its presence, but the policy has not secured for it a stake comparable to other established or emerging great powers due to geopolitical constraints and aspects inherent in its strategic culture. Its go-it-alone approach has earned it little by way of global influence or regional presence in Central-South Asia. In the meantime, the rise of China and its perceived strategy to encircle India have erected new barriers for Delhi's emergence as a great power.

Meanwhile, the U.S. strategic presence in the region leaves much to be desired. The United States is withdrawing from Afghanistan without major political or military gains from the conflict that has cost it and its partners a fortune in lives and money. The future of its military infrastructure and relationships with countries in Central-South Asia is a big unknown, with regional partners equating the U.S. military pullout with its waning commitment to support the regional economic and security order. To help unlock their strategic potentials, Delhi and Washington should join forces in Central Asia. Despite promising signs, doing this is not going to be easy, given political barriers, institutional mistrust, and past grievances of both nations.

Whether India will rely on its own resources—to be expanded with the rise of its economy—or seek a partnership with the United States, or both, remains to be seen. What is clear is that India's geopolitical role is set to expand considerably in the next years and decades. Such expansion will generate concerns for neighbors and distant actors alike, creating conditions for instability despite benefits of cooperation between them and India as one of the largest economies. Washington and Delhi should ensure they stay engaged in Central Asia and enhance their positions amid a power struggle unfolding between outside powers in this increasingly critical part of the world by mitigating Afghanistan-Pakistan (Af-Pak) security challenges; shaping Iran's geopolitical role; fostering Sino-Indian cooperation; and exploiting grand strategies and regional connectivity initiatives of other actors. This they should achieve as part of a strategic partnership that makes Central Asia its major pillar. Until then, neither Delhi nor Washington is likely to succeed.

The role of the U.S. military in the process, either as part of the U.S.-Indian strategic partnership or as part of U.S. overall efforts to cultivate such a partnership with Delhi, will be critical. Rightly so, because a lot is at stake for the United States in this seemingly remote backwater. As an area of growing external rivalries, the region of Central-South Asia is a source of both traditional and nontraditional security threats to U.S. national interests, be that in the political, military, or even economic domain. From interstate conflicts to transnational terrorism and from Russia's attempts to reestablish geopolitical control to China's efforts to achieve economic dominance, the region is a focal point of intersecting challenges and opportunities that the U.S. military should be better positioned to address and leverage in support of U.S. national interests. Pursuing those objectives as part of U.S. economic, political, and military efforts would help India unlock its strategic potential and assist Washington in unlocking that of its own.

I: INDIA'S DELAYED ARRIVAL AND ASPIRATIONS IN CENTRAL ASIA

India's impressive economic growth over the last 2 1/2 decades has brought India's role and interests in the world and Central-South Asia to the forefront of global politics and statecraft. Importantly, it has put India into a comparative perspective with China, another aspiring Asian great power poised to stiffen competition for resources and influence among established and emerging powers alike. Both are resource-hungry and rapidly emerging powers seeking a new place and role in the global economic and security orders. Both are also strategic rivals and consider their immediate neighborhood of Central Asia of growing strategic importance to their grand strategies. Just as China has viewed its westward expansion into Central Asia imperative to acquire resources and project power globally, so has India deemed its northward push into the region essential to nurture and expand its strategic potential as an aspiring global power.

For now, China has outperformed India in Central Asia on all counts, securing the region as a key resource base and platform for power projection. In 2012, India launched the "Connect Central Asia" policy to shore up its regional presence, but the policy has not secured for it a regional stake comparable to other established or emerging great powers. Select elements of India's strategic culture and a number of geopolitical constraints explain the country's lagging position behind China and its constrained role in the world and the Central-South Asian region.

Indian authorities have not fully discarded the legacy of India's nonalignment ideology and the role of this ideology in the country's foreign policy, which has only recently started assuming a more assertive posture that takes national interests, not ideology, as its baseline. They are also coping with India's lost orientation following the dissolution of the Soviet Union and are still defining India's global and regional visions and roles. Finally, they continue to be preoccupied with India's domestic challenges impeding more focused external policies and engagement. Together, these factors make India's external direction slow and unclear, its policies less targeted and sustainable. Unless Indian elites address these issues, enhancing India's global and regional position will prove a harder enterprise for the country that sees itself as an aspiring great power.

The same goes regarding India's ability to address a number of geopolitical constraints, which are products of geography, history, and India's relations with neighbors. The existence of a disconnected broader region of Central-South Asia is chief among them. Despite the collapse of the Soviet Union, which set Central Asian countries free to pursue links with neighbors in South Asia, and the U.S. military presence in Eurasia's heartland for the first time in history, which opened a connection between Central and South Asia, both regions remain largely disconnected from within and without. They are also the areas that are least integrated with the global economy and whose constituent units have tense relations, impeding trade and economic development. India does not share a border with Central Asia and has to rely on the relatively isolated Iran, unstable Afghanistan, and airlifts from Delhi and Dubai to trade with the region.

India further confronts what I call the "quicksands" of Afghanistan and Pakistan—a status quo of now chronic instability and security risks in the region that sustain the disconnection between Central and South

Asia and limit significantly regional economic development and India's strategic reach in and beyond both regions. The instability in Afghanistan and Delhi's tensions with Islamabad prevent India from establishing a direct and expanded link with Central Asia. Prospects of civil war, separatism, and disintegration in Afghanistan and Pakistan present additional challenges for India's efforts to connect with the region. Delhi views its ascendance in Central and South Asia as critical for its plans to become a great power. But it cannot achieve this status without mitigating the challenges facing Afghanistan and Pakistan and addressing its tensions with Islamabad and Beijing, especially as they concern the transit role of Kashmir.

Besides geopolitical constraints involving Pakistan, China, and Afghanistan, India also faces a highly competitive role played by China, with which it fought a war and still has unresolved border disputes. Delhi is concerned about China's emergence as a neighboring great power and related impact on India's global position in a potentially reformatted global economic and security order. It is also especially wary about Beijing's perceived encirclement strategy to keep India tied to South Asia and the growing role of Pakistan as Beijing's platform to deny Delhi an active regional role. The legacy of the 1962 border war and Sino-Indian border disputes reinforce this perception, impeding the development of transcontinental links via Central-South Asia (especially through Xingjian and Kashmir) with the participation of India, China, and Pakistan.

Despite its growing capabilities, India is starting its ascendance to the ranks of great powers from a lower economic base and at a slower pace compared to China and other Southeast Asian economies. It is the last major Asian economy to join the Asian economic renaissance—a courtesy of its leadership during the Cold War that considered the success of others "as largely irrelevant to its [India's] own future" and favored the "continuation of existing policies" as a sign of the country's autonomous role in world affairs.[1] Only in the 1990s did Indian authorities start advancing economic reforms with a focus on privatization, a process that continues to this day and still faces enormous challenges. Unlike other Asian economies, India embarked on reforms after the collapse of the Soviet Union and amid expanding U.S.-led global economic integration and the risk of "increasing marginalization" if India failed to reform. The urgency of economic reforms was so grave that it prompted one commentator to compare India's shift of economic gears with India's second independence, only this time "from a rapacious and domineering state."[2] As a result, India is only now starting to enjoy the fruits of its economic reforms.

Those factors help explain why India's trade with Central Asia was just $U.S.1.24 billion (B) in 2013-14 compared to China's at U.S.$50B or why, despite friendly relations, India's model of development is not appealing and the regional states do not view India as a major counterbalancing force in their external strategies just yet. To beef up its foreign policy and to change the status quo in the region, in 2012, India launched the "Connect Central Asia" policy to link South and Central Asia via energy, trade, and transit corridors—a constrained connectivity initiative of an aspiring great power (India) with an agenda similar to the mission of the poorly executed New Silk Road Strategy (NSRS) of a relatively declining power (the United States). If successful, the policy would create a sustainable economic corridor between Central and

South Asia, helping India project its presence in both regions and achieve a great power status sooner.

Be it as it may, India's "connect" policy faces tough competition, especially from the Russia-led Eurasian Economic Union (EEU) and the China-led Silk-Road Economic Belt ("belt"). This is despite India's promising areas of engagement with Central Asian republics in areas as diverse as energy resources exploitation, nuclear energy cooperation, transit infrastructure development, trade facilitation, military and defense collaboration, development of space exploration related programs, and external strategic balancing. The policy is also conceptually concentrated on and limited to reconstruction and integration of Afghanistan with Central and South Asia, while lacking financial and diplomatic resources for this task. It further ignores India's strategic need—no less geopolitically significant as the internal and external integration of Afghanistan—to resolve tensions with Pakistan and China and capitalize on their roles in reconnecting the regions and unlocking India's strategic potential. Therefore, despite its projected rise as a great power in the long term, India's south-north "push" in the next decade will be weaker and less assertive than China's east-west "pull." Bureaucratic politics at home and instability in the immediate neighborhood will keep India largely confined to South Asia and the Indian Ocean.

The regional position of the United States, too, faces major challenges. Washington and its allies are in the process of withdrawing from Afghanistan—a development that evokes concerns of unmanaged instability in the region and U.S. disregard of Central-South Asia at the time when new security challenges and opportunities are emerging, which could either undermine or reinforce U.S. global standing and its long-term capacity to remain a global power. What is more, it is a U.S.-Indian strategic cooperation that is increasingly in demand to ensure that both Washington and Delhi succeed in addressing these security threats, capitalizing on these opportunities, and achieving their objectives in Central-South Asia and the world at large.

Currently, the U.S. strategic presence in the region and achievements as part of the NSRS leave much to be desired. By 2016, the United States is likely to withdraw from Afghanistan on schedule, without major political or military gains from the conflict that has cost it and its partners a fortune in lives and money. The future of its military infrastructure, including bases, agreements, and relationships with countries in Central-South Asia, is a big unknown, not only for its allies in the North Atlantic Treaty Organization (NATO), but also partners in the region. The latter equate the U.S. military pullout with its waning commitment to support the regional economic and security order, to which Washington contributed but which it is now struggling to shape.

The fate of the U.S. NSRS is no longer as important as it was originally promoted to be, and not only because of its own deficiencies in the form of limited funding, lacking commitment, and inactive leadership. Other actors are increasingly implementing similar concepts in practice and with a lot of success, enabling them to advance their national interests while advancing regional development. Beneficial as it is in some ways for the region, the implementation of similar connectivity initiatives by other actors without an effective and committed participation by the United States in its own initiative, let alone other projects, risks marginalizing the U.S. regional role of promoting globalization and development as key pillars of its national security strategy.

In the meantime, efforts of established powers to maintain their positions in the face of growing presence of emerging powers complicate the already complex mosaic of relationships in the region, surrounded as it is by four nuclear powers. Central and South Asia are experiencing instability but are reconnecting after decades of isolation. Iran is poised to become a key regional economic integrator after decades of isolation imposed by the West. China and India are rising as great powers and seeking to advance their interests with growing force. Russia seeks to reassert its traditional geopolitical role in the post-Soviet space. In short, parts of Eurasia are seeing a reconfiguration of their geopolitical maps—the processes the U.S. military should be ready to shape to advance U.S. national interests in the rapidly changing strategic environment.

Indeed, the continent is undergoing rapid power shifts brought about by the rise of new powers and assertiveness of established ones, while confronting threats from nonstate actors like the Islamic State (IS) seeking to advance their agenda at all cost. But while the United States is present militarily in Central-South Asia, it is barely present economically and plans to disengage militarily at exactly the time when its overall presence is of growing importance to its ability to pursue new military missions in Eurasia and remain a relevant global player. India's situation is similar, but only to an extent: it, too, is hardly present in Central Asia, but has a strong desire to set a deeper footprint as a way to achieve a status of great power. What they have in common is their lack of strategic presence in the region yet combined potential to be significant actors in Central-South Asia.

To help unlock their strategic potentials globally and regionally, Washington and India should partner in Central Asia. Doing this is not going to be easy. Both states share political impediments, institutional mistrust, and a number of grievances. India views it increasingly important to carve out for itself an independent role reflecting its ascendance and confirming its status as a rising power and has traditionally positioned itself as an autonomous actor unwilling to be a "pawn" of other powers. Washington, in turn, may not want to commit to a strategic partnership that does not reflect a bigger voice and role by the United States, among other reasons. Yet, both recognize the growing importance of aligning their views and joining forces in tackling emerging challenges and opportunities. India's go-it-alone approach has earned it little by way of global or regional presence, while the U.S. do-it-alone approach has undermined its influence worldwide. In the meantime, the rise of China is yet to erect barriers to Delhi's emergence as a great power and Washington's ability to retain its status as the strongest actor.

Latest developments in the U.S.-Indian relations indicate that both actors are strengthening their strategic cooperation without compromising each other's vital interests. In 2015, Delhi and Washington concluded another nuclear energy cooperation deal, opening the door for Delhi to import U.S. technologies critical for India's status as an ascending technological power in sectors other than information technology (IT). This they did despite India not being a member of the Non-Proliferation Treaty (NPT). The agreement comes a decade after the parties concluded the first nuclear deal, seen as a "paradigm shift" in India's geopolitical code, with Delhi feeling increasingly comfortable partnering with the United States to counter China's advances in areas viewed as lying within India's periphery.[3] Both are pursuing similar connectivity strategies toward Central-South Asia. And both are increasing bilateral cooperation—in large part to even the scales of global power flows—as they confront a rising China in the Pacific and Indian oceans and now across Eurasia.

The United States would benefit from a much stronger strategic presence by India in the heart of Eurasia. The former is a relatively declining power and a democracy that can facilitate India's ascendance to a great power status quicker and with less obstacles. The latter is a rapidly emerging power and a democracy with an increasingly entrepreneurial and growing middle class, which can assist the United States in ensuring a peaceful global power transition from "the West to the rest." Both should also be less concerned about competing with each other for strategic influence in Central Asia. The United States is far from the continent, does not wield influence in the region comparable to Russia or China, and would welcome India's stronger presence to ensure that neither Beijing nor Moscow dominates the region. India, in turn, is closer to Central Asia but is a democracy, lacks any imperial legacy of expansion into Central Asia and, importantly, does not enjoy a strategic presence and perceived intent to dominate the broader region—all while facing Russia's and China's opposition to its regional advances.

A strategic partnership with a focus on Central Asia between the United States and India should be premised on joint and unilateral actions aimed at mitigating Af-Pak security challenges to facilitate India's linkages with Central Asia; shaping the geopolitical role of Iran to advance U.S. and Indian interests in the broader region; fostering Sino-Indian cooperation to enable India's linkages with China, Pakistan, and Central Asia via Kashmir and Xingjian; and exploiting grand strategies and connectivity initiatives of other actors in Central-South Asia, such as Russia and China, to advance their own policies. Importantly, the U.S.-Indian partnership should not constrain Sino-Indian cooperation where it advances India's position in Central-South Asia. Beijing plays a growing role in the evolution and stability of Afghanistan and Pakistan, and its interest in developing links between the Middle East and Central-South Asia could significantly facilitate U.S. and Indian regional policy if Delhi and Washington play their regional cards right.

The United States will not be able to achieve the previous objectives without relying on its military. From mitigating security challenges in the Af-Pak region, given the planned military withdrawal from Afghanistan and reformatting its missions in light of potential U.S.-Iranian détente, to promoting confidence building and shared security and crises management approaches with China and India and protection of vast and expanding economic infrastructure throughout Eurasia, the support of the U.S. military to the overall U.S. strategy in the region is going to be critical for U.S. general efforts to unlock India's strategic potential across all four objectives. The earlier Washington engages Delhi in the region, the easier it will be for Washington to shape India's emergence as a great power to support U.S. regional and global security agenda.

India's rise is undeniable and set to expand its global reach and Central Asia's connectivity with South Asia. With time, Central Asian states may find it imperative to deflect the growing pressure from the east (China) and north (Russia) by cultivating closer ties with the south (India), as they look for a secure way of connecting with the world as independent units. India's potentially transformational regional engagement is there to help. But Delhi is currently a latecomer to the region's 21st century "Great Game" and faces tough competition from Russia and China, making it critical for Delhi to partner with Washington to bolster its strategic presence in the region. Until then, India's arrival in Central Asia will remain delayed for the foreseeable future, making it harder for Delhi to turn its regional aspirations into sustainable long-term gains. Regardless of whether Delhi and Washington enter a strategic partnership in the region, the United States should better understand and help India address select elements of its strategic culture and geopolitical constraints that are impeding its efforts to connect with the region.

II: INDIA'S STRATEGIC CULTURE AND GEOPOLITICAL CONSTRAINTS

Undefined Role, Unclear Direction.

A number of elements of India's strategic culture explain Delhi's limited engagement with Central Asia. India's legacy of nonalignment, lost orientation, and focus on internal agenda following the collapse of the Soviet Union, and relatively passive foreign policy have rendered its global role poorly defined and executed and, ultimately, unfulfilled. As a result, India has not pursued a clear direction, projected an appealing normative model of development, or exclusively relied on its national interests as the guiding principle of its foreign policy. But India recognizes these issues and, given its projected emergence as an influential global player, has sought to adjust its foreign policy in light of unfolding global dynamics and its national interests.

While India officially discarded the ideological underpinnings of its nonalignment after the end of the Cold War, its foreign policy continues to reflect and project related principles. India refuses to participate in power-balancing schemes or serve perceived agendas of other power(s). This makes it difficult for Delhi to utilize the leverage of its potential allies and partners in areas of the world where its role is limited compared to its strategic needs. In Central Asia, where its geopolitical presence is insignificant but its needs and future role are potentially enormous, aligning with other actors would benefit Delhi. However, India tries to avoid becoming a "pawn" in the perceived U.S. chess game of containing China and Russia in the heart of Eurasia.

In a way, Delhi finds it difficult to part with the legacy of its nonalignment model because it has simply not yet developed a new one. Pursuing a foreign policy reflecting elements of nonalignment allows it to preserve some sense of direction in the otherwise less controlled geopolitical environment. Ironically, the end of India's nonalignment *de facto* came not with the end of the Cold War, but with it. After Indian and Chinese troops clashed along the border in 1962 and after Washington and Beijing reached détente in 1972, India was compelled—despite its promoted status of nonalignment—to align itself with the Soviet Union to balance the Sino-Pakistani-U.S. axis,[4] while leaving room for a strategic autonomy vis-à-vis Moscow and retaining popularity in the third world. The demise of the Soviet Union shattered Delhi's perception of its strategic environment and its role as the leader of the nonalignment movement.

India's foreign policy was disoriented, while its domestic policy was preoccupied with development challenges caused by the rupture of economic links with the Soviet Union. Former Indian Ambassador to the United States Abid Husain, an economist, described India as a tiger in a cage, a potentially powerful yet hesitant actor: "When the cage is open, the tiger would show its real strength. The cage is now open but the tiger refuses to come out of the cage."[5] Focusing on domestic challenges after the collapse of the Soviet Union distracted India from pursuing an active foreign policy, which has lacked assertiveness and only recently started operating in geopolitical codes rather than ideological frames of nonalignment. Its foreign policy also focused on the immediate region of South Asia, while treating Central Asia as Moscow's periphery and neglecting China's growing influence in the region. This explains India's "gap between ambitions and capabilities" in the region that persists to this day.[6]

India's inward-looking focus and preoccupation with domestic agenda have further prevented it from the creation of a normative and appealing development model and pursuit of a larger objective beyond its borders, even as the country has been gaining economic strength. Leading Indian intellectuals at the 1919 Paris Peace Conference reportedly envisioned an independent India that would not project power outward or seek to dominate other states. This left little room for "a global objective in Indian strategic culture," contributing to the emergence of a nonalignment tradition that bore the hallmarks of "sovereignty and autarky" viewed as "independence from the rest of the world, rather than as a particular way of engaging others."[7] As an expert on India's security policy, Achin Vanaik observes that India's foreign policy has not had a "real strategic vision or well-developed strategic sense, being too prone to a moralistic form of posturing as a substitute for pursuing hard-headed and self-serving foreign policy perspective." The lack of recognition of the importance of power accumulation and power politicking, Vanaik argues, allowed external forces to "repeatedly invade and defeat India over the centuries."[8]

India's proclivity to wait rather than act is another aspect attributed to its strategy, culture, and foreign policy. Rodney W. Jones, an expert on South Asia, for instance, argues that India is determined "to wait the opponent out" rather than engage it in an effort to produce "a pragmatic compromise"—a position reflecting India's "profound sense of entitlement, superiority, and presumed deeper knowledge about the correlation of forces that distinguish India's strategic culture."[9] The result is a rising India that remains a spectator with a "wait-and-watch-and-hope-for-the-best" approach[10] rather than an active actor using every opportunity to enhance its presence in the region of growing importance for its ascendance as a great power. Unlike China and the EEU, whose "normative power" is based on the power of active example, India's is based on the power of "passive example." Delhi has an expectation that other actors will follow India without it having to provide any inducements.[11] But it has not yet succeeded in leveraging effectively its rich popular identity, maturing politics, and economic progress.[12]

India's anti-colonial struggle, nonalignment tradition, and unfulfilled economic potential have discouraged Delhi from actively promoting itself as a model, let alone imposing its vision of development on other countries. In Central Asia, it seeks to lead the way for regional republics—without a model—in areas such as democratic and secular development, building of multi-ethnic societies, and economic modernization. But even such an unimposing approach comes with a price. India's political influence in the region is very weak. Delhi does not support regional opposition groups, considers the evolution of the regional states and societies through the lens of Islamic currents, supports a multi-ethnic federal system rather than unitary nation-state building premised on a dominant ethnic group, and has failed to demonstrate convincingly the fruits of its economic modernization given its widespread poverty and legacy of its cast system.[13]

India's economic growth in high digits over the years (7.5 percent this year) and tensions with China and Pakistan nevertheless have prompted Delhi to pursue a more assertive foreign policy. India's nuclear test in 1998 demonstrated Delhi's more assertive foreign policy posture not just vis-à-vis nuclear-armed Pakistan and China, with which it has fought several wars (1948-49, 1965, 1971 with Pakistan, and 1962 with China), but also in relation to the United States and others with which it seeks equal great power relations. The test revealed Delhi's technological prowess and, importantly, its intent to show off its capabilities as a rising actor ready to assume a geopolitical role no longer constrained by the Cold War.[14] As former Indian Foreign Affairs Minister Pranab Mukherjee explained, India's intensified engagement with South Asia and increasingly other areas of the world is "the beginning of the reassertion of India's historically benign and stabilizing role in these regions, premised on the commerce of ideas and goods."[15]

According to Indian strategist C. Raja Mohan, India's grand strategy is preoccupied with three concentric geographic circles: an immediate neighborhood where India seeks primacy; an "extended neighborhood" covering the rest of Asia and Indian Ocean littoral where it seeks to counterbalance other powers and prevent them from undercutting its interests; and the rest of the world where India sees itself as a great power capable of playing a key global role.[16] On a regional level, and specific to Central Asia, India gradually has transitioned to a more active formulation of its policy toward the region, as well. It started with its "Look North" and "Look West" policies toward Central-West Asia, initiating its "Connect Central Asia" policy in 2012. The "connect" policy seeks not to "**look**" north but to "**connect**" the north, reflecting an emphasis on action rather than observation. India intends to pursue a more direct and impactful engagement with Central Asia by making Afghanistan the hub of interregional reconnection.

But intent is not a capability, and India is yet to overcome a number of deficiencies related to its strategic culture and foreign policy, as well as major connectivity and geopolitical constraints, before it can enjoy strategic presence in Central Asia even remotely comparable to great powers.

Constrained Interregional Connectivity.

India faces a number of connectivity constraints limiting its presence in Central Asia: the fact of a disconnected Central and South Asia as a broader region; India's lack of a contiguous border with Central Asia; significant geographical barriers; India's unresolved political tensions with Pakistan and China; and interstate rivalries within Central and South Asia. The instability of Afghanistan and Pakistan is another major factor, which is addressed in a separate section.

India and the rest of South Asia had interacted with Central Asia for centuries. But the division of spheres of influence between the tsarist Russia and Great Britain in the 19th century, the creation of the Soviet Union in the early-20th century, and the partition of India in 1947 sealed the borders between the two regions, leaving them largely disconnected—a status quo that in a significant way persists to this day.[17] India found itself largely cut off from Central Asia during the Cold War, despite maintaining deeper ties with Central Asian republics compared to other powers due to its friendly relations with the Soviet Union until the latter's collapse in 1991. The demise of the Soviet Union allowed Central Asian states to pursue links with South Asian neighbors as independent units, but the constituent states of both regions failed to develop strong relations with each other or neighbors. The regional republics were preoccupied with day-to-day survival and consolidation as nation-states, guarding their newly found independence and perceiving countries in South Asia as less stable, less developed, and more risky partners, instead seeking ties with Russia, Turkey, European partners, and the United States, among other actors.

Civil war in Tajikistan in the 1900s and instability in Afghanistan from the 1900s onwards further impeded the development of any linkages between Central and South Asian countries. Both Tajikistan and Afghanistan represented and still represent connection points between Central and South Asia, making it crucial for their elites and counterparts in neighboring states to address any ongoing security concerns centered on separatism, cross-border militancy, and Islamist agenda. The instability in Tajikistan and Afghanistan facilitated cooperation between India and Central Asian states aimed at stabilizing the security situations in both countries. Today, this cooperation is in the works, given India's concerns about the regional instability impacting its projected rise. But the planned exit of coalition forces from Afghanistan and the potential departure of Washington may delay and constrain India's efforts to reconnect with the region, even if it would stimulate a need for Delhi's deeper engagement with Central Asia.

The U.S. military invasion of Afghanistan after September 11, 2001 (9/11) has been a transformational development for the broader region. But the reconnection process it set in motion may stall, if not reverse, when Washington and its coalition allies pull out of Afghanistan. The presence of U.S. forces in the heart of Eurasia for the first time in history led to the creation of the Northern Distribution Network as a corridor running supplies across Central Asia to coalition forces in Afghanistan. It has also encouraged intraregional and interregional collaboration on economic, political, and security matters, as concerned parties have searched for a solution to the conflict in Afghanistan and better prospects for their development. But the longevity and aftermath of this reconnection process as driven by the military presence of coalition forces is a big question mark after 2016. The end of this process and India's do-it-alone approach in the region portend complications for Delhi's efforts to reconnect Central and South Asia via a more stable, developed, and integrated Afghanistan.

Unlike China, Russia, or Iran, India also lacks a contiguous border with Central Asia, which makes the pursuit of trade and transit links with the region a much harder enterprise. India has to rely on the relatively isolated Iran, unstable Afghanistan, and airlifts from Delhi and Dubai to trade with the region. Geographic barriers presented by the Himalayan mountain range complicate its reach north as well, making the development and advancement of India's links with China's Xingjian and Tibet a challenging, though not impossible, task on technical grounds. (Interestingly, China has boosted its connectivity infrastructure within Tibet and Xingjian provinces for economic and military purposes. But India has lagged behind in similar efforts in its northern areas,[18] including in Aksai Chin—a region adjacent to Jammu and Kashmir that China administers but India disputes. Beijing uses Aksai Chin to connect Xingjian and Tibet via its national highway.) Moreover, the Indian subcontinent is rather "self-contained," with a harsh terrain and an ocean isolating it on all sides and making it more of an island than a territory integrated with Eurasia.[19] Robert Kaplan correctly points out the limiting factor of geography on India's global and regional ambitions: "India is a regional power to the degree that it is entrapped by this geography; it is a potential great power to the degree that it can move beyond it."[20]

India's rivalries and tensions with Pakistan and China, along with perceived security threats from Bangladesh and Nepal, present a major obstacle for India to connect with Central Asia, while "robbing India of vital political energy" to project power across Eurasia.[21] Pakistan, India's rival in South Asia ever since India's partition in 1947, denies India overland access to Central Asia, forcing it to rely on airlifts, sealanes, and overland links via Iran and then Afghanistan to trade with the region.[22] Meanwhile, Sino-Indian disputes, including over Aksai Chin and parts of Arunachal and Himachal Pradesh, continue to strain the relationship between the two powers that once fought a war over contended borders. India's borders with Pakistan and China are currently either sealed or partially open at few crossings, impeding transcontinental trade and transit.

Rivalries and tensions within Central and South Asia among constituent states further impair interregional connectivity by hampering linkages within and between respective regions. In Central Asia, stricken by border, water, and ethnic disputes (especially in the Fergana Valley shared by Tajikistan, Uzbekistan, and Kyrgyzstan), a relatively isolated Uzbekistan and relatively open Kazakhstan struggle for regional primacy. In South Asia, riddled by insurgencies and militancy, the nuclear-armed Pakistan and India jockey for regional influence, as well.

India has a lot to accomplish in Central-South Asia, as both regions remain largely disconnected from within and without and represent the areas that are least integrated with the global economy. This makes the involvement of great and emerging powers in the regions a source of opportunities and challenges, advancing the regions' external integration but with a potential cost to the sovereignty and independence of the constituent countries. India is one such power that is projected to be an economic engine driving the reconnection of the regions, provided it properly addresses the connectivity constraints and Af-Pak challenges limiting its regional engagement.

Quicksands of Afghanistan and Pakistan.

To a significant degree, India's limited presence in Central Asia is a factor of the war-torn Afghanistan and volatile security situation in Pakistan. The unfinished war in Afghanistan and the planned withdrawal of coalition forces from the country by 2016 threatens further instability in Central-South Asia and reversal of any modest connectivity gains that India and the regions have attained at a big cost. Meanwhile, economic and security challenges in Pakistan may swell considerably, threatening the country's disintegration and, along with Afghanistan, inviting the involvement of state and nonstate actors and exacerbating the already fragile political and security fabric of Central-South Asia. India is yet to realize that its success in Afghanistan and Central Asia are hardly possible without its significantly improved ties with Pakistan and China—the countries playing a major geopolitical and geo-economic role in Afghanistan and broader Central-South Asia. Absent that, India will remain trapped in the quicksands of the Af-Pak region, unable to project a meaningful geopolitical role in either Central-South Asia or beyond.

After the withdrawal of coalition forces, Afghanistan may experience disintegration due to a potential flare-up of civil war or major sectarian conflict that could engulf neighboring Iran, Pakistan, and Central Asian states. Expansion of the conflict beyond the borders of Afghanistan and Pakistan could bring about systemic perturbations, if not for the global security system then certainly for the Eurasian continent or its geopolitical subsystems. This, in part, explains Delhi's focus on reconstruction of Afghanistan and reconnection of Central and South Asia via Afghanistan as the interregional node of trade, energy, and transit connecting the two regions. This thinking in India's foreign policy is significant. It goes beyond Delhi's traditional and still strong obsession with Pakistan to encompass its responsible reconstruction role and a forward-looking policy vis-à-vis Central Asia despite, indeed because of, the war in Afghanistan.

India is concerned about Islamabad's and Beijing's efforts to draw Kabul into their fold, especially after coalition troops leave Afghanistan. A Taliban-ruled Afghanistan once provided Islamabad with a platform for its strategic depth strategy that fed India's sense of insecurity.[23] China's growing presence in Afghanistan may yet deal a blow to India's plans seeking to establish its own strategic foothold in the country. India's rivalry with Pakistan and China over Kabul is important to India for securing Afghanistan as the interregional link advancing Delhi's agenda. But in the current shape, this process binds India to a two-front struggle with neighbors, sapping its resources and accentuating heavily the competitive rather than collaborative dynamics in its relationships with Islamabad and Beijing. Delhi should reformulate its engagement in Afghanistan and Central Asia by pursuing a cooperative accommodation with Pakistan and China focused more on collaboration than competition or disengagement. Achieving either outcome would be impossible without China and Pakistan changing their postures vis-à-vis India, as well. Currently, such will is weak in all three capitals—a dangerous status quo, considering Pakistan's growing strategic role for China in promoting transcontinental development via Central and South Asia without the participation of India.

A more positive cooperation involving China, Pakistan, and India would be hard to achieve without addressing the dispute over Kashmir and the latter's role as a trilateral and transcontinental connector. The division of Greater Kashmir into Jammu and Kashmir and Azad Jammu Kashmir, administered by India and Pakistan respectively, and the establishment of a Line of Actual Control in 1949 over which China, India, and Pakistan fought several wars in the 1960s, 1970s, and 1900s undermined any prospects of expanded transcontinental links.[24] The status quo in Kashmir constrains Indo-Pakistan and Sino-Indian ties; prevents expanded transit and trade between Central, South, and East Asia; and could threaten an open interstate conflict. The three countries view their control over parts of Greater Kashmir as a source of military advantage in a possible escalation of conflict, reinforcing a perceived need to control the evolution of the region in ways that hamper rather than facilitate trade via this territory. Instead, they could cultivate regional stability by fostering the independence of a unified Kashmir. Since this position seems untenable at this time, the parties should pursue other possible arrangements.

Besides tensions with Pakistan and China, as well as instability in Afghanistan, India also has to grapple with potentially game-changing regional ramifications of a volatile Pakistan. The issue of Pakistan's territorial integrity has haunted its elites since the country's creation in 1947, when it was carved out of India and saw a part of its territory now known as Bangladesh secede in 1971 in a bloody war. Nearly 4 decades since, Islamabad has struggled to ensure domestic security amid militant attacks, some of which bear the hallmarks of separatism. Pakistani authorities fight Lashkar-i-Jhangvi and Pakistani Taliban in Baluchistan, federally administered tribal areas, and Khyber Pakhtunkhwa, besides struggling to maintain their hold on Kashmir. The funding of militant groups against Soviet forces and India has ricocheted against Pakistan. In 2014 alone, militant and terrorist attacks against the state and minorities left approximately 5,500 people dead, including civilians, security forces, and insurgents/terrorists.[25] The chronic instability could threaten popular unrest or disintegration of the nuclear-armed state, producing consequences with which India would not be able to cope alone and that would exacerbate the instability in Afghanistan and further complicate India's efforts to reconnect with Central Asia.

The less opportunities Pakistan has for economic development, the higher the risks of its disintegration and negative ramifications for neighbors in the region, especially India. Delhi should pursue a policy aimed at not only reconstruction and reconnection of Afghanistan but also incorporation of Pakistan in the regional economic system, with and without urgently needed political arrangements necessary to decrease its tensions with Islamabad. Ensuring a responsible and constructive role by Pakistan in domestic and regional security should inform India's political, economic, and security dialogue and engagement with China as Pakistan's main partner. India's success as an aspiring great power requires a projection of power and a search for resources to buttress such power in order to address geopolitical dilemmas centered on Afghanistan, Pakistan, and China. India's search for genuine cooperation with Pakistan and China would accelerate the stabilization, reconstruction, and integration of Afghanistan into Eurasia's fabric of commerce, while unleashing India's strategic potential. Until then, the quicksands of the Af-Pak region will continue constraining Delhi's global and regional reach.

China's Head Start and Encirclement of India.

As if facing volatile Afghanistan and Pakistan were not enough, Delhi also has to contend with ambitions of China—the only other Asian power challenging India's long-term rise and search for resources, routes, markets, and bases in Central Asia. China had an earlier start in the race, initiating economic reforms in the 1978-80s and unleashing a wave of economic development unprecedented in history. It has also pursued more skillfully its perceived strategy of encirclement of India, while Delhi has lagged behind with its own policies countering China and unleashing its own potential, including in Central Asia. China's more successful use of partnerships as part of a more homogenous and centralized foreign policy and sharing of a borderline with Central Asia has ensured China a more expansive influence in the region.

China's advantages over India notwithstanding, India has a number of attributes potentially making it the largest economy and a major strategic partner for Central Asia in the long term: India's democratic system, relatively young and increasingly mobile population, entrepreneurial spirit of its growing middle class, and the wide use of English language, among others.

China undertook economic reforms about a decade earlier than India. As an early starter, it achieved the status of the largest economy in 2014. With an economy worth $U.S.17.6 trillion (T) compared to the U.S. economy worth U.S.$17.4T (based on purchasing power parity), China enjoys the largest economic presence in Central Asia. India, on the other hand, began "waking up" only in the early-1990s after the collapse of the Soviet Union and the end of the Cold War forced it to reconsider its domestic and foreign policy in light of the transformed external environment. India then embarked on privatization of its economy (which continues to this day) and pursuit of a foreign policy increasingly guided by national interests rather than an ideology of nonalignment that drew on perceived impractical ideals. Still, India continues to lag behind China in economic development, especially in infrastructure investment, manufacturing, and education. Similar to China, it struggles with pervasive corruption, challenges of privatization, as well as sectarianism and separatism in a number of states. Its democracy in the short term is no match for China's authoritarian system that "can make things happen" on command. In India, locals say "development takes place in spite of, rarely because of, government."[26] While India's society has remained traditionally strong, its state has remained relatively weak.[27]

But unlike China and Russia, and similar to democracies like the United States, India's government is unable to direct businesses to serve its geopolitical objectives in a given region with precision and foresight widely attributed to Beijing and Moscow. India does not have effective multilateral economic platforms in Central Asia, its businesses do not get considerable support of the government,[28] and it does not make businesses a major part of its grand strategy. The companies led or controlled by China and Russia, on the other hand, have been relatively successful in acquiring assets and shoring up the countries' influence in Central Asia and elsewhere. India's increased economic and political presence in the region is thus a factor of more expanded collaboration between the Indian business community and the government.[29] In the short term, India's approach of building its economy from the bottom up is a disadvantage to its economic position vis-à-vis China, which pursues a top-down development approach.[30]

Beijing has also outsmarted India in geopolitics, leveraging alliances and partnerships as part of its relatively coherent and centralized foreign policy. In the race for resources and power, China views India as a rival and is seen as encircling China along its flanks and tying it down to the confines of South Asia. To that purpose, China has allegedly used its growing economic and security partnerships with Nepal, Bhutan, Bangladesh, and Myanmar to India's northeast; Sri-Lanka, Thailand, and Malaysia to India's southeast; Pakistan and Afghanistan to India's northwest; and Central Asian states to India's northwest.[31] India's border war with China in 1962 and political tensions with Beijing have only underscored India's perception about Beijing's perceived attempts to outflank it.[32] China's alleged suggestion to the U.S. Pacific Fleet in 2009 to recognize the Indian Ocean as part of a Chinese sphere of influence have further reinforced India's perception of Beijing's disregard of Delhi's ambitions and security interests.[33]

While China has relied significantly on Pakistan to outdo India in South Asia, it feels no need to rely on any other actor(s) to do so in adjoining Central Asia. Beijing already enjoys a relative edge in terms of economic presence and influence, relegating India to the status of a constrained, middle-ranked power that does not even boast a rapidly growing settlement in and emigration of its nationals to Central Asia. Nevertheless, China had to partner with Russia on global and regional issues to advance its position in Central Asia. Beijing and Moscow are members of the Shanghai Cooperation Organization (SCO) opposing the U.S. perceived unilateralism in world affairs. By recognizing Russia's predominant security role in Central Asia and proposing a free trade zone between the Moscow-led Eurasian Union and the Beijing-led "belt," China seeks an enhanced position in the Central Asian region without unnerving Moscow (just yet).

Despite this "marriage of convenience," Beijing's ambitions threaten Russia's long-term interests in Central Asia. This creates an opening for Delhi and Washington to exploit Moscow's concerns in order to promote their influence in Central Asia, including by seeking an SCO membership for India. India's membership in the SCO would enhance India's ties with regional states and enable Delhi to shape its competitive relationship with China on more collaborative terms.[34] India applied for membership in 2014 but continues to confront resistance. Moscow supports India's membership to dilute China's presence in the region, but China objects to it for fear of losing a share of its growing influence in Central Asia. The SCO's rules and consensus among members that candidates must be signatories to the NPT and not have conflicts with members is another obstacle.[35] Besides meeting these conditions, established or presumed, Indian elites would have to address their concerns about policy constraints, a loss of strategic autonomy, and association with authoritarian regimes that could come with India's membership in the organization.[36]

Besides seeking the SCO membership and partnering with other powers to counter China's perceived encirclement strategy, India has also been making its own advances in countries lying on China's periphery. In Southeast Asia, it has acquired assets in the economies of Myanmar, Nepal, Sri Lanka, Bhutan, Thailand, and Vietnam. In Central Asia, it has cooperated with the regional states and Afghanistan against the Taliban in the 1990s and after 9/11, imported arms from Uzbekistan, sought rights to the use of the Ayni airbase in Tajikistan, assisted with training of the Kazakh Caspian fleet, opened a mountain biomedical research center, and participated in counterterrorism exercises with Kyrgyz special forces in Kyrgyzstan.[37] To counter China's military encirclement, India is modernizing its own military capabilities, enhancing its power projection capacity, building logistics and transit infrastructure in the north, and conducting military exercises with its distant and nearby partners.[38] It has also sought the creation of a shared region-wide energy market in Central-South Asia, in part by pursuing the Central Asia-South Asia 1000 (CASA-1000) project and Turkmenistan-Afghanistan-Pakistan-India (TAPI) initiative to bring electricity and gas from Central to South Asia. After it lost to China the Petro

Kazakhstan deal in 2005, India further proposed a Pan Asian grid backed by a U.S.$22.4B commitment, but the idea went off the radar after a reshuffle in its Petroleum Ministry.[39]

Despite their tense and competitive relations, India and China have been improving their ties over the last 3 decades, with an increase in bilateral trade being a vivid demonstration of this development that may yet translate into more friendly ties. The complementarity of their economies—India is a leader in services production, while China is a leader in manufacturing—adds another positive spin on the bilateral relationship that is already enjoying a bilateral trade turnover of U.S.$71B (though India's trade deficit with China of U.S.$47B is a thorn). Chinese Premier Wen Jibao, for instance, remarked that India and China had conflicts only briefly in the 2,000-year history of exchanges, and the bilateral ties have been 99.9 percent friendly.[40] As Kaplan explains, "India's rivalry with China is not like the one with Pakistan at all: it is more abstract, less emotional, and (far more significantly) less volatile. And it is a rivalry with no real history behind it."[41] Furthermore, as former chief of the Indian navy Arun Prakash argues:

> Both China and India have certain common threads running through their history, and this helps rationalize their contemporary behaviour as nation-states. Both are ancient civilizational entities with a very strong religious-cum-cultural underpinning which explains the enduring nature of their mores and traditions. About 500 years ago, these two nations were so prosperous that, between between them, they contributed over 50 per cent of the world's GDP. Both have experienced invasions; for most part from Central Asia, but from the fifteenth century onwards, increas-

ingly across their shores. While the overland invaders, whether they prevailed or were defeated, became assimilated into the two resilient cultures, it was the Western interlopers, coming by sea, who inflicted subjugation and humiliation on both nations. This has created a deep national urge never to allow a re-run of history, and to regain past glory as early as possible.[42]

To be sure, both China and India face seemingly intractable border disputes in Jammu and Kashmir, Himachal Pradesh, and Arunachal Pradesh. At least 323 border incursions were reported in 2011 along the Sino-Indian border, highlighting the extent of the dispute over the borderline and the risk of interstate collision. As of 2012, China maintained 400,000 troops in Chengdu and Lanzhou, while India deployed 235,000 troops along its borders with Tibet.[43] China, in addition to Pakistan and other nations in South Asia, have been "sustaining several insurgent groups" in India.[44] Meanwhile, India is refining its "two-front war" doctrine vis-à-vis China and Pakistan, while still perceiving China as a threat. Despite subsequently improved rhetoric, India had viewed Beijing as its number one threat right before the country's 1998 nuclear test, which it justified in part by pointing to the perceived threat posed by China. As India's Prime Minister Atal Bihari Vajpayee explained in a letter to U.S. President Bill Clinton:[45]

> I have been deeply concerned at the deteriorating security environment, especially the nuclear environment, faced by India for some years past. We have an overt nuclear weapons state on our borders, a state which committed armed aggression against India in 1962. Although our relations with that country have improved in the last decade or so, an atmosphere of distrust persists mainly due to the unresolved border problem. To

add to the distrust, that country has materially helped another neighbor of ours to become a covert nuclear weapons state. At the hands of this bitter neighbor we have suffered three aggressions in the last 50 years. And for the last ten years we have been the victim of unremitting terrorism and militancy sponsored by it in several parts of our country, especially Punjab and Jammu & Kashmir.[46]

But both increasingly realize the importance of a positive dynamic in their relationship as two rising powers and are working to resolve their disputes. They mitigated tensions over Sikhim, opened up the Nathula Pass across the Tibet-Uttar Pradesh for trade, and may yet facilitate India's links with Central Asia via Aksai Chin and Xingjian.[47] They also agreed to expand trade by opening up the Lipu-Lekh Pass linking India, China, and Nepal. Hu Shih, China's former ambassador to the United States, commented on prospects of cooperation this way: "India conquered and dominated China culturally for 20 centuries [through Buddhism] without ever having to send a single soldier across her border." Will it do so again, but economically?[48]

Moreover, one could interpret the perceived strategies of encirclement by China and counterstrategy by India as run-of-the-course expansions of economic and military power and influence of both rapidly growing Asian nations, which are hungry for resources, energy, and opportunities and are coincidentally seeking to secure them by extending respective influence in the same regions. This interpretation, however, has to withstand the test of time, geography, and projected geopolitical dynamics currently positioning the two powers as rivals and forcing India to rethink its grand strategy of pursuing a strictly autonomous role in world affairs. As an authoritarian, relatively homogenous, and economically wealthier state, China has pursued a more effective, national interest-based foreign policy that engages elements of power balancing and partnerships with allies and promotes China's model of development worldwide, including in Central Asia. China and India also view resource-rich Central Asia as a platform of competition more than cooperation between the two giants. Beijing seeks to deny India a leverage over resource deliveries bound for China from and via the Middle East and Central Asia-South-Asia,[49] even if India's participation in China's transcontinental projects could open new doors for Beijing.

If they are not to leave India behind, Indian elites need to think in geopolitical and strategic codes more expansive than India's immediate environs, projecting the country's future against global dynamics in light of India's potential status of a global power in the long term. They should also ponder India's contribution to the world and its appeal as a developing and, later, a developed nation. At this time, China's model of development—no matter how appealing, poorly-designed, or promoted—is one asset India does not yet possess, not in Central or South Asia. This helps explain India's relatively limited presence in both regions and its recent push at projecting its influence in the greater region as part of its "Connect Central Asia" policy.

III: INDIA'S SEARCH FOR A WAY OUT: PROSPECTS AND LIMITATIONS

Delhi's "Connect Central Asia" Policy.

To mitigate challenges impeding its reconnection with Central Asia and to expand its regional presence, India launched the "Connect Central Asia" policy in 2012. The policy has enabled it to advance links with Central Asia via Afghanistan but is yet to secure a regional strategic presence comparable to other powers. Addressing deficiencies of the policy and ensuring the implementation of TAPI, construction of a proposed U.S.$40B gas pipeline from Russia to India via Central Asia and China, and expansion of CASA-1000 and the International North-South Transit Corridor (INSTC), among other projects, is critical. It would help India, the sixth-largest energy consumer, become a magnet for regional resources and a source of exports for the entire post-Soviet space, while unleashing its strategic potential as an emerging great power.

According to India's former Minister of State for External Affairs Shri Ahamed, the "connect" policy is based on proactive political, economic, and people-to-people engagement with Central-South Asian countries, both individually and collectively. As part of the policy, India announced plans to set up 14 flight links with Central Asian states; develop local IT, energy, banking, and pharmaceutical industries; and build energy infrastructure and e-networks linking Central and South Asia.[50] Delhi also planned on establishing a new Comprehensive Economic Cooperation Agreement with the region.[51] The policy, which seeks to promote new and expand existing connections with the region, should contribute to India's foreign policy objectives, both regional and global. As former secretary at India's Ministry of External Affairs Rajiv Sikri put it:

> India would like to encourage the development of stable and secular regimes in Central Asia, lest weakened, unstable states with centrifugal tendencies become bases for terrorist, separatist, and fundamentalist elements, which could link up with counterparts in Afghanistan and Pakistan. . . . India seeks to have a firm foothold and exercise influence in Central Asia along with other great powers so that this strategically located region does not become an area dominated by forces inimical or hostile to India's interests. . . . Aspiring to be an influential global power, India has to be a player in the unfolding 'Great Game' in Central Asia, in an equal footing with the other major players like the United States, Russia, and China if it is to successfully protect its vital national interests in Central Asia.[52]

But Delhi is far from attaining an "equal footing" on par with Moscow or Beijing. The "connect" policy does not pursue an explicit military or security initiative focused on Central Asia. It further lacks effective multilateral and bilateral frameworks of cooperation with the region. This does not mean India is not trying. India is a member of the Conference on Interaction and Confidence Building Measures—the only organization having India and Central Asian states as members.[53] It would also like the regional states to become members of the South Asian Association for Regional Cooperation (SAARC) in either the existing or a separate framework, while seeking membership in the SCO. Delhi has also been cultivating security and military ties with Central Asian states. It has found in Tashkent its major arms supplier and tried but failed to establish an airbase at Ayni in Tajikistan, which it helped refurbish. It also assisted with training of the Kazakh Caspian fleet, opened a mountain biomedical research center, and participated in counterterrorism exercises with Kyrgyz Special Forces in Kyrgyzstan for the first time in 2015. Finally, India has cooperated with Central Asian states on stabilization of Afghanistan. But Russia's traditional and China's emerging security presence and role present major obstacles to India's efforts to establish its own security and military foothold in the region.

While Russia has beat India in the race for military bases and geopolitical influence, China has outdone it in the race for markets and geo-economics presence—a sphere of activity that forms an overwhelming component of India's "Connect Central Asia" policy. In 2013, China acquired an 8.4 percent stake in Kashagan oil field, which Kazakhstan had previously promised to sell to India for the same amount of U.S.$5B.The deal was one of 20 bilateral agreements worth U.S.$30B. India's Oil and Natural Gas Corporation (ONGC) is currently attempting to buy out ConocoPhilips' 8.4 percent stake in the Kashagan oil field.[54] Beijing also outperformed Delhi in securing rights to develop the Galkynysh gas field in Turkmenistan and is launching a pipeline in 2009 to supply Turkmen gas to China. Beijing and Astana further opened a second pipeline to bring Kazakh oil to China. Finally, China has invested far more resources in the region's transit infrastructure, while boosting its economic presence in Afghanistan where it more visibly competes with India.

Before India launched its "connect" policy, it established intergovernmental commissions for trade, economic, scientific, and technical cooperation with the regional republics.[55] But it has attained only limited gains. In 2014, 2 years after it launched its "connect" policy, India's trade with the region stood at approximately U.S.$1.24B compared to China's at U.S.$50B. The same year, India's exports and imports to the Central Asian region constituted less than 2 percent and 1.5 percent of its overall exports and imports on average, respectively. India largely imports cotton, raw materials, zinc, uranium, iron, steel, and dried fruits from the region, while exporting tea, textiles, leather, rice, pharmaceuticals, IT technologies, and chemical products.[56] This is why implementing TAPI, constructing the gas pipeline from Russia to India, expanding CASA-1000, and broadening INSTC is the priority for the "connect" policy. After all, cooperation between India and Central Asian republics is especially promising in the energy and transit development sphere, given the lack of transit infrastructure between Central and South Asia and India's limited presence in the region's energy market. As of 2011, India was not even on the list of top 10 countries involved in the oil and gas exploitation in Central Asia.[57]

The TAPI gas pipeline project, estimated to cost U.S.$10B, would reduce India's dependence on energy imports from the Middle East, facilitate improvement in Indo-Pakistani ties, and advance development and integration of Afghanistan with the broader region. It would also challenge EEU's energy interests in the Caspian, undermine Russia's grip over the region's energy exports, and help the Central Asian states diversify their ties to balance China. But the TAPI participants have failed to kick-start the project for financial and security reasons ever since the idea originally came up in 1995-96, despite making significant progress recently. The parties plan to start the construction of the pipeline in 2015 with a view to make it operational by 2020.

Construction of the gas pipeline extending from Russia via Central Asia to China and then to India is another proposal India has been advancing to facilitate its resource acquisition strategy in Russia and Central Asia. The pipeline would transport gas from Russia via Turkmenistan, Uzbekistan, and Kazakhstan to Kashgar in China's Xingjian, before crossing into India via Ladakh or Himachal Pradesh.[58] However, the project faces technical, financial, and geopolitical challenges and is subject to the resolution of territorial disputes and mitigation of Sino-Russian and Sino-Indian rivalries for Eurasian resources and trade. The interest of Russia and Central Asian states in the project, Russia's overall leverage vis-à-vis China, and Beijing's own interest in expanding its gas imports offer some prospects for implementation of this proposal.

The CASA-1000, in turn, is already operational but could be expanded to increase electricity exports to Pakistan and India. As TAPI, it would bring India and Pakistan closer together, if not politically then at least economically, to ensure they are on a sustainable development path and are not a significant source of security threats for neighbors. As an Indian observer noted, any electricity lines extending from Central Asia via the Wakhan corridor would probably traverse areas of Kashmir occupied by Pakistan and India, making the Line of Control increasingly irrelevant as parties expand electricity and other types of trade.[59] Besides financial challenges, the project confronts a limited electricity production capacity of Tajikistan and Kyrgyzstan in the short to medium term, as well as opposition of downstream Uzbekistan, Kazakhstan, and Turkmenistan to required dam construction in the two upstream countries that the downstream states fear would deprive them of even more water resources. Delhi also has to face an emerging role of China as its rival for electricity imports, as Beijing seeks to import Central Asian electricity to develop its lagging regions of Xingjian and Tibet. India and China should approach the issue of importing the region's electricity with an eye to mitigate the potential for interstate conflict, in the same way they should in the case of oil and gas imports from Central Asia.

The INSTC, launched by Russia, India, and Iran in 2000, is already partly operational but is yet to be finalized and should be expanded to include Central Asian states and Afghanistan, and potentially Pakistan and China, if India wants its reconnection with Central Asia to occur sooner rather than later. Kazakhstan, Tajikistan, and Kyrgyzstan in Central Asia, as well as Armenia and Azerbaijan in the South Caucasus, among other countries, joined the project and are considering ways to tap into the initiative advancing sea, rail, and land connections between India, Iran, Russia, and Europe. India helped with the construction of the Zaranj-Delaram and Zaranj-Milak roads linking Iran and Afghanistan but should link and expand them as part of the INSTC.[60]

The isolation of Iran due to the standoff between Tehran and Washington and India's tensions with China and Pakistan—besides challenges presented by geography or nonstate actors—have impeded these four geopolitically monumental projects. This is despite the argument that implementation of the initiatives would help address these very challenges. For now, Delhi is leveraging its economic ties with partner countries to advance these projects and to increase its overall presence in Central Asia irrespective of their implementation.

In Kyrgyzstan and Tajikistan, India is involved in hydropower development initiatives. It also plans to open an Indian-Central Asia university and a military hospital in the two countries, respectively. In Uzbekistan, India is present in the pharmaceuticals, IT, construction, energy, and mining sectors. In Kazakhstan, its firms are actively involved in coal, oil, and uranium industries. From 2009-14, India imported more than 3,500 tons of uranium from Kazakhstan. Delhi is also interested in building a gas pipeline from southern Kazakhstan and has boosted its engagement with Iran, given a potential détente between Tehran and the West.[61] But instability in Pakistan and Afghanistan and Tehran's unresolved standoff with the West have hampered its efforts to import energy resources via proposed TAPI and Iran-Pakistan-India gas pipelines.[62]

But India's main focus as part of the "connect" policy has remained on Afghanistan.[63] Over the last decade, it has invested U.S.$2B in the country's infrastructure and won rights to develop the Hajigak and other deposits (Afghanistan's overall deposits are estimated at U.S.$1 to 3T). India has even deployed paramilitary forces in Afghanistan for the first time to protect its assets and personnel.[64] Delhi committed $U.S.100 million (M) to develop the Iranian port at Chabahar and spent U.S.$136M to connect the port with Afghanistan's Ring Road. The port would enable Delhi to access Central Asian markets without relying on Pakistan and position it favorably vis-à-vis China, which helped build a rival Pakistani port at Gwadar linking China and the Persian Gulf. The Chabahar port forms a part of the INSTC, designed to expand a south-north vector of the transcontinental trade.[65] As of 2013, about 60 percent of Afghanistan's trade passed via a road network built by India and Iran linking Chabahar and Bandar Abbas ports with Central Asia.[66] Besides its funding for roads, railways, medical facilities, and power networks, India helped Afghanistan become an SAARC member to facilitate its development and regional integration.

India is interested in seeing Central Asian states as members of the SAARC, as well, and more generally seeks to extend its cooperation frameworks with other states to Central Asia and the Middle East, including the South Asian Free Trade Area (SAFTA), the Twelfth Bay of Bengal Initiative for Multisectoral Technical Economic and Cooperation, and the India-Association for Southeast Asian Nations deal.[67] It would also like to see regional republics, especially Tajikistan, Uzbekistan, and Kazakhstan, join SAFTA, which has Afghanistan as a member and Iran and China as observers.[68] SAFTA is an agreement reached in 2004 that created a free trade zone for an area covering Bangladesh, Bhutan, India, Maldives, Nepal, Pakistan, and Sri Lanka with the combined population of 1.8 billion people. The parties plan to reduce customs duties of all traded goods to zero by 2016. India and Pakistan acceded in 2009, while Afghanistan acceded in 2011.

Granted, the "connect" policy is more active than the "Look North" policy, has more funding, and focuses on Afghanistan's reconstruction as the key to reconnection of Central and South Asia. It also promises a more profound involvement of India in the region's future in the years and decades ahead. But it is not without its own challenges. It has no explicit linkage to the resolution of Sino-Indian and Indo-Pakistani disputes. It does not factor in the role of China, whose economic expansion can help India remove corks and unleash its expansion. It also lacks effective multilateral economic and political mechanisms of engagement with Central Asia in economic and, importantly, security spheres. The projects that underpin the policy lack financial means compared to China's "belt" initiative. Delhi also does not issue checks as easily as Beijing does and is not fully in the business of securing strategic assets and investments in the region at all cost. Finally, the policy lacks alignment with similar strategies of other actors, especially the United States, which pursues a similar agenda in the region. India thus should adjust the way of doing business in Central Asia by exploring its prospects of competition and collaboration with other actors in Central-South Asia to facilitate its regional position.

External Competition and Cooperation.

India's relationships with external actors in Central Asia exhibit competitive and collaborative dynamics. But its ties with Pakistan and China are not only more competitive and strained, but also potentially conflict prone if the parties fail to reach some type of accommodation in the long run.

The Indo-Pakistani rivalry in South Asia is as old as the two countries themselves. Ever since the independence and partition of India in 1947, Delhi and Islamabad have jockeyed for territorial integrity, independence, and geopolitical positioning by fighting four wars (1948-49, 1965, 1971, and 1999); acquiring nuclear weapons; sponsoring militant and secessionist groups; engaging in trade spats and restrictions; and seeking financial, diplomatic, and military assistance from other powers. Similarly, the Sino-Indian competition for power and influence in Asia has been ongoing since the 1940s when China and India were established as modern states.[69] India and China, a close backer of Pakistan, went to war with each other over disputed territories in 1962.

India's rivalry with both actors is assuming new dimensions, with major implications for Eurasian stability. First, China's rise as a global power and its expanding maritime and continental presence in areas surrounding India is turning Pakistan, and potentially Iran, into a primary component of Beijing's transcontinental resources acquisition strategy and a platform of power projection along the rimlands and across Eurasia. This could aggravate the Indo-Pakistani and Sino-Indian rivalry if Delhi fails to cultivate cooperative dynamics in its relationships with Islamabad and Beijing. Second, India's emergence as a global power and its own expanding maritime and continental presence are prompting Pakistan and China to search for allies to keep India in check, potentially exacerbating the Indo-Pakistani and Sino-Indian rivalry, as well.

The projected expansion of India's and China's activities in Central Asia and reconnection of Central and South Asia add a twist to the Indo-Pakistani and Sino-Indian relationships, to the role of Pakistan in Beijing's policy to keep India in check, and to the role of Pakistan and China in Delhi's efforts to unlock India's strategic potential in Central and South Asia. Were they to ease tensions, India, Pakistan, and China could benefit enormously from connectivity initiatives being advanced throughout Eurasia. Empires that used to rule over territories now occupied by all three states had once already extracted major geo-economic advantages from the Silk Road by facilitating transcontinental circulation of goods and ideas.[70] Resolving Sino-Indian and Indo-Pakistani border disputes as major land disputes afflicting Asia would open up many of the broader region's areas that have remained locked since the middle of the 20th century.[71]

Implementing and expanding TAPI, CASA-1000, Iran-Pakistan-India (IPI), INSTC, and the China-Pakistan Economic Corridor would demonstrate a great potential of trilateral cooperation in promoting national and interregional development. The China-Pakistan Economic Corridor (CPEC), in its current design, passes entirely through Pakistan and its governed area in Kashmir, which India disputes. Turkmenistan and Kazakhstan may join the IPI in the future. India could import gas from Russia and/or Central Asia via Xingjian and the disputed Aksai Chin. It could also build links with Central Asia via Gilgit, Wakhan, and Srinagar-Muzaffarabad corridors running via Kashmir if the parties resolve their disputes.[72]

Currently, China and Pakistan are hesitant to involve India in energy pipeline projects given their competition for energy resources. Both signed a U.S.$7B deal to extend the Iran-Pakistan pipeline northward to China without involving India.[73] But were they to include India, Pakistan and China could earn transit fees, enhance their trade with India and the region, and gain leverage over Delhi.[74] China and India already share experience, pursuing joint projects, including in the energy sphere. They bid jointly for oil stakes and for exploring oil fields in Colombia, Kazakhstan, Sudan, and Iran. They also agreed to consider exploring jointly gas fields in Indonesia and Australia.[75] Currently, the parties are considering launching an India-China oil consortium in Central Asia.[76] Both countries further conducted joint search-and-rescue operations off the Shanghai coast in 2003 and anti-terrorism exercises in China in 2007 when Indian forces set foot on Chinese territory for the first time since the 1962 war. More generally, they have held regular military exercises and exchanged delegations on all issues.[77] Besides, China does not perceive India as a strong competitor in Central Asia,[78] providing room for enhanced cooperation between the two countries in the broader region in the short term.

Addressing lingering issues with Pakistan and China is key for Delhi's and Beijing's less restrained role in South and Central Asia. While China and India are bound to be strategic rivals in both regions in the long term, they have to find ways to cooperate to avoid conflict and maximize their development potential. Neither China nor India alone is capable of reformatting the geopolitical field of Eurasia in order to benefit fully from any Eurasian economic development. But both countries can mitigate and prevent interstate frictions and conflicts in the continent by emphasizing collaborative dynamics in their ultimately competitive bilateral relationship. India needs to push north, while China is already seeking connections with the Middle East via Iran, Afghanistan, Pakistan, and, potentially India to complement east-west links and diversify and expand its reach.

India's positive relationship with Iran is also instrumental for advancing transcontinental connections and India's strategic presence in Central Asia, especially if Iran and the United States start improving ties. Iran borders energy-rich Turkmenistan and volatile Pakistan, and shares a cultural heritage with Tajikistan and India, presenting a number of avenues of cooperation with Delhi in trade, energy, and security spheres focusing on Central Asia. Delhi has already boosted its ties with Iran in response to the start of nuclear talks in 2014 but is yet to tap into Iran's geopolitical position and capacity to advance its interests in Central Asia. India especially eyes Iran's energy resources and transit potential but is increasingly concerned about China's advances in Iran. Beijing has invested U.S.$120B into Iran's energy sector over the last 4-5 years. As of 2011, Iran was China's second-largest oil supplier, accounting for 14 percent of China's oil imports.[79] An expanding cooperation with Iran enables India to enhance trade with Central Asia and keep China's ambitions centered on Iran and Central Asia in check. Central Asian states are, in turn, interested in engaging Iran but cannot do so in a scalable way for fear of distancing themselves from Washington and other actors. Iran, in its turn, would like to expand its role in Central-South Asia, while diversifying its energy exports. Iran and Kazakhstan seek participation in the Uzbekistan-Turkmenistan-Iran-Oman corridor, while Kazakhstan, Iran, and Turkmenistan already operate a rail line linking the three countries.[80] Iran's regional advances would ultimately compete with India's, but the two would benefit from aligning their policies to promote their otherwise constrained engagement in the region in the short to medium term.

India's ties with Russia are also helpful to India's plans in Central Asia but, like any other aspiring hegemon, Russia does not always like to share. Moscow would be happy, in the words of Vladimir Jirinovsky, to "wash the boots" in the Indian Ocean, expand the INSTC, and see Delhi become a member of the SCO to balance China in Central Asia. Moscow already seeks to participate in north-south projects backed by the West, such as TAPI and CASA-1000, to at least shape, if not control, the reconnection of Central and South Asia.[81] Russia would also like India's assistance in diluting U.S. global influence in the quest for a more multipolar international order.[82] But it would not like India to have a strong role in Central Asia, as its pressure on Dushanbe to prevent Delhi from getting its first-ever overseas military base at Ayni has demonstrated. Some reports indicate that Moscow's decision to block India's access to the base in Tajikistan was to "punish" Delhi for trying to diversify its weapons suppliers.[83] But the alleged punishment may be one aspect of Russia's general opposition to India's military role in Central Asia. Delhi's best bet is to cultivate ties with Russia, exploit Moscow's concerns about China's rise, and leverage Russia's regional position to advance its interests in Central Asia.

But it is Delhi's evolving relationship with Washington that portends major global and regional implications, especially if it assumes a strategic character with a focus on Central Asia. India is already a counterbalancer against China, a role the United States welcomes and a role that is autonomous rather than subordinate to U.S. global interests. As Mohan argues, "India has never waited for American permission to balance [against] China," and has been doing so since China's invasion of Tibet.[84] Increasingly, Delhi and Washington face a need to work together in Central, South, and East Asia. Washington has also sought to promote integration of the broader region as part of the NSRS, which is overlapping with India's "connect" policy. Both policies focus on reconstruction, integration, and positioning of Afghanistan as the integrating bridge between Central and South Asia, seeking to create a common energy market in Central-South Asia by advancing TAPI, CASA-1000, and other initiatives, while providing a platform for the Central Asian and South Asian republics to pursue southward vectors of development. Not least important are shared challenges facing both the NSRS and the "connect" policy, which should prompt Delhi and Washington to cooperate in advancing their regional influence.

But Delhi and Washington have legitimate reservations about engaging one another in Central Asia. India continues to oppose U.S. perceived unilateralism in a quest for a more just and multipolar order, is not willing to be perceived as a U.S. pawn, is concerned about disrupting its ties with other emerging and established powers seeking to keep Washington in check, and is generally held back by the legacy of its nonalignment foreign policy tradition. The United States, on the other hand, risks undermining its ties with China and Russia and escalating external rivalries in Central Asia. It could also facilitate dynamics that would deprive the EEU of potential energy imports, which will be increasingly diverted to Southeast Asia, and potentially exacerbate conflict-prone competition over influence and resources in the Central Asian region.

Examining India's relationships with external actors sheds light on challenges and opportunities of India's current and future engagement with Central Asia. But no less important are India's relationships with individual countries of Central Asia, as they are only starting to shape up.

Regional Views and Areas of Engagement.

Potential areas of engagement between India and Central Asia are profound, especially because India lacks effective and meaningful bilateral and multilateral mechanisms of engagement with Central Asia.[85] But the parties need to do a lot of homework to realize this engagement.

Central Asian states welcome India's regional presence for a number of reasons. India is not only an aspiring great power—itself a major consideration for the regional countries located in its immediate periphery. It is still a developing nation that has shown great achievements in economic growth and does not seek to impose its will or way of development on others. It also remains a democratic and secular nation, respectful of diversity and numerous ethnic and religious groups that are more optimistic about India's future than they were about 2 decades ago when India embarked on economic reforms. Central Asians do believe India can offer a model of development in the short term that could compete with Russia's or China's. But they consider India's successful economic development as a source of inspiration for the regional economies and populations that are eager to connect with South Asian neighbors and the world.

As a potential great power, India offers regional republics an opportunity to ride the wave of development faster and counter ambitions of grandeur emanating from Moscow, Beijing, Ankara, or Tehran. Central Asian states cannot disregard this opportunity, not when outside parties seek to test or project their models of development in Central Asia.[86] Provided the regional states start building links with one of the fastest growing economies (and potentially the largest) early on, they can tap into opportunities projected to spring up from these connections. Deepening the economic linkages would enable the regional economies to integrate with their neighbors to the south. Delhi's more substantive economic engagement would translate into geopolitical leverage that the regional countries can use to balance external pressures from actors that once hosted the largest empires and continue to be active geopolitical actors on the Eurasian chessboard. India also offers the regional states a platform of technological cooperation as an important attribute of modernization for the regional economies. India is actively positioning itself as a technology power capable of overcoming geopolitical constraints in the broader region and is seeking to build for itself a regional presence worthy of a modern, 21st century power. In Central Asia, it has opened several technology centers, largely in in the information sector.[87]

Central Asian countries and India share not only history, but also culture and religion going back hundreds of years. This cultural connection remained alive somewhat even during the Soviet times when India was one of only four countries allowed to have a consulate in the Soviet Central Asia.[88] India is also home to the second largest population of Muslims, many of whom practice a Sufi tradition of Islam predominantly practiced in Central Asia. What is more, Muslim countries of Central Asia prefer a stronger presence of India rather than Pakistan, being concerned as they are about fundamentalist and terrorist movements from Pakistan seeking to undermine the political and social fabric of the regional states and societies. They view Islamabad with suspicion, given its role in aiding the Mujahedeen in the fight against the Soviet Union and Pakistan's perceived inaction vis-à-vis terrorist groups that originated in Central Asia but are now based in Pakistan, such as the Islamic Movement of Uzbekistan. The regional states also share the memory of China's control of part of the region in the past[89] and welcome India's role to offset Beijing's current economic expansion in the region. Yet, India's perception of the regional states sympathizing with a Sharia-ruled Pakistan, its preoccupation with rivalries with Pakistan and China, and its inward looking policies since the 1990s have misframed its policy, making it assume a "belated, slow, and half-hearted stand" toward Central Asia.[90]

As India continues its rise and Central Asian states continue their quest for external integration, more room for collaboration between the parties is expected in the coming years. While the stabilization, development, and regional integration of Afghanistan will remain a key pillar of their cooperation, other promising areas include energy, trade, and transit facilitation; defense collaboration; space programs development; and external balancing. Of all these areas, energy cooperation is a major priority for India, whose oil and gas imports are forecast to increase to 90 percent and 40 percent, respectively, by 2020 in order to meet the country's projected energy demand.[91]

Turkmenistan is a major prospective energy partner for India, while Delhi is an emerging vector of balancing against China's growing presence in Turkmenistan's energy market and Russia's traditionally strong security role in the Caspian. Delhi views Turkmenistan as a critical component of its energy import diversification strategy, especially after India's ONGC-Mittal acquired 30 percent of shares of two oil sites in the Caspian in 2007 but withdrew from the projects in 2010 citing "exploratory failures."[92] The construction of TAPI would be a major milestone, not only for India and Turkmenistan but also for Afghanistan and Pakistan, which suffer from lack of security and face potential disintegration. India would have a source of energy resources from Central Asia and a major infrastructural connection with the region, which would entail a much more expanded and dynamic economic presence of India in the region's market. Turkmenistan would diversify its energy exports, this time not just vis-à-vis Russia but also China, which has recently replaced Russia as the largest Turkmen gas importer. Afghanistan and Pakistan, in turn, would tap into the region's energy inflows to power their economies and help ensure sustainable economic development that promotes regional stability.

Uzbekistan and India, in turn, have a lot of ground to cover in the areas of gas and uranium exploration, defense cooperation, transit infrastructure development, and external balancing. India is already engaged in the joint exploration and exploitation of gas fields and construction of liquid gas and oil factories in Uzbekistan,[93] but more room exists for cooperation in this sphere. India looks to Uzbekistan to diversify sources of uranium imports and to boost cooperation in arms production—critical ingredients for India's expanding nuclear capability and emerging indigenous arms industry. Delhi also considers Uzbekistan the key platform of transit connections in the region. Uzbekistan borders all Central Asian states and Afghanistan and has major rail and highways passing through its territory.[94] But instability in Afghanistan and measures by the Uzbek regime to close down the country to its own citizens, neighbors, and the world has hampered the bilateral transit infrastructure development cooperation, even if Tashkent and Delhi could pursue a meaningful collaboration to integrate Afghanistan with Central-South Asia. Despite its central location, Uzbekistan has largely kept itself aloof from the rest of Central Asia and preferred bilateral rather than multilateral engagement.[95] This partly explains why India has increasingly viewed Kazakhstan as its platform in Central Asia.[96] Tashkent would like to see India alongside other Southeast Asian states to balance Russia and China. But until the parties expand their political engagement, this prospect is far off.

As Central Asia's largest economy, Kazakhstan presents significant opportunities for cooperation with India in oil and gas exploration, uranium production, trade facilitation, space programs development, and strategic balancing. Access to oil, gas, and uranium is critical for India, which depends heavily on uninterrupted imports of these strategic resources from overseas. But India's energy presence is confined to "ancillary activities"[97] and is yet to expand. In 2009-10, Videsh Limited got a 25 percent stake in the Satpayev block for the first time since 1995, when India started seeking a foothold in Kazakhstan's energy sector.[98] India also looks to expand trade with Kazakhstan, which accounts for 75 percent of India's trade with the region. But the creation of the EEU and the potential creation of a free trade zone between the EEU and China could complicate India's efforts unless it manages to develop a similar arrangement.

Space development is emerging as a major pillar of India-Kazakh cooperation. India is interested in creating a landing space, utilizing a radar complex, and using Kazakhstan's territory for launching its remote sensing satellite rockets. Astana, in turn, is interested in India's space development program to develop that of its own, eyeing the expertise of a technology center in Bangalore and the possibility of expanding its satellite launching clients.[99] Especially crucial to Astana is a potential platform provided by India to keep Kazakhstan's two large neighbors in check as part of its praised multivector policy. Russia and China impact Kazakhstan's evolution as an independent and resource-rich state, which is the largest (by territory) in the post-Soviet space after Russia, making Delhi a key partner for Kazakhstan. Both countries are also rapidly developing and emerging as leaders in their regions, seeking a status of modern and technological powers.

India's cooperation with Kyrgyzstan and Tajikistan is largely about access to hydro resources, trade facilitation, defense cooperation, and gaining military presence in the two countries. The hydro potential of Kyrgyzstan and Tajikistan is immense and presents major opportunities for India's engagement in the production and export of electricity as part of CASA-1000. Trade facilitation and education are other prospective areas, but they remain limited for the time being.

Defense and military cooperation, however, is emerging as an important pillar, with Delhi seeking to establish military presence and stronger military-to-military ties with both countries. In Kyrgyzstan, India eyes a torpedo production plant and torpedo testing site in northern Kyrgyzstan,[100] besides opening a biomedical military research facility and holding Special Forces exercises in 2015 for the first time. In Tajikistan, it refurbished a military hospital at Farkor military facility and a runway at Ayni military base, cooperating with Dushanbe against the Taliban in the 1900s and since 2001. Tajikistan borders Afghanistan, China, and Pakistan along the Wakhan corridor, making the country important to India's military contingencies.[101]

India's bet on Kyrgyzstan and Tajikistan to advance its military presence in Central Asia is not coincidental. Kyrgyzstan and Tajikistan are weaker and have less room for autonomous foreign policy compared to their larger and richer neighbors. But even here, India's moves could be too little, too late due to Russia's substantial military presence in both countries. Still, Kyrgyzstan and Tajikistan remain interested in boosting security and economic collaboration with India, considering it an evolving partner in their external balancing strategies. Moreover, India has viewed Tajikistan as a platform to enhance the bilateral framework of its engagement with Central Asian republics,[102] while looking to Kazakhstan to boost its overall regional presence.

Prospects for engagement between India and Central Asian republics clearly exist. But until they address instability in Afghanistan and, to a lesser degree, volatility in Pakistan, their reconnection is going to be slow, ineffective, and painful. As they ponder these hard questions, they should necessarily factor in the challenges and opportunities presented by the two unstable countries and minimize the likely fallout in the form of expanded flows of refugees, terrorism, and narcotrafficking due to reconnection of Central and South Asia. Cultivating relations is itself a difficult undertaking when you are starting from a low base, but doing so in the conditions of instability in neighboring countries that are linking Central and South Asia is a significant challenge, requiring India to search for allies in the region to shore up its regional presence.

Search for Partners as Strategic Imperative.

Identifying and working with partner(s) in advancing its presence in Central Asia is a strategic imperative for India if it wants to establish a more prominent presence in the region similar to other great powers. Given numerous limitations on its ambitions, Delhi should advance cooperation with its perceived rivals, such as China and Pakistan, and actors that it views as its prospective strategic partners, such as the United States. However, the premise of this work is that a closer partnership with Washington in particular would help India overcome these limitations and better achieve its agenda. Building such a partnership and advancing its regional presence is a challenge for Delhi. It faces major constraints inherent in its strategic culture and foreign policies, hampering its ability to promote influence globally and regionally. But failure to accomplish related tasks risks delaying significantly India's rise or making India fall further behind.

As a potential great power, India is losing out to all other great powers in the region, including Russia, China, the EEU, and the United States. It also finds itself struggling to compete with actors as middle- and small-ranked as Turkey, the United Arab Emirates, Pakistan, and Iran. Perceptions of its potential emergence as a great power notwithstanding, India remains a middle-ranked power confronting a whole range of internal challenges that will likely preoccupy it for at least the next 1 to 2 decades before it can turn itself into a major global power. To make this transition faster, India needs partners to help it enhance its presence in Central Asia and beyond. Otherwise, its expansion as a great power will proceed primarily along the rimlands of Eurasia.

Unlike Russia, China, and the United States, India starts from a lower base of engagement with the region since the collapse of the Soviet Union. This is despite India's relatively substantial economic presence in Afghanistan, where Pakistan's involvement has long defined India's policies that are only now starting to be proactive, go beyond India's obsession with Pakistan, and reflect Delhi's appreciation of Afghanistan's role as an interregional integrator. It also faces major economic and social development constraints that keep its energies focused on internal rather than regional challenges and opportunities, which are both restricting and facilitating its expansion as a great power. Moreover, India does not have as much cash as China to throw around and has a lot to do at home first. Partnering with other actors on a strategic level could open doors to financial, diplomatic, and military support India needs to achieve its objectives.

India also does not border Central Asia directly, having to lean on the unstable Afghanistan and somewhat isolated Iran to access the region. The instability in Afghanistan and uncertainty surrounding the potential for détente between Iran and the West leave many doors closed for Delhi, both in terms of its bilateral relations with these countries and in terms of its engagement with Central Asian republics. Were these two issues addressed, India would be in a more geopolitically favorable position to advance its "connect" policy and escape the confines of South Asia. Partnering with a democrat great power that not only has access to Central Asia but is also capable of influencing outcomes related to the evolution of Afghanistan and Iran is key to India's active foreign policy toward Central-South Asia and the Greater Middle East.

Further, India has strained ties with nuclear-armed Pakistan and unresolved border tensions with nuclear-armed China—the neighboring countries with which it has fought a number of wars. Positioned between Pakistan and China in power rankings, India finds it hard to change the status quo alone. Having a partner capable of influencing China and Pakistan could give India a chance to attempt a change in its relations with the two neighbors. Otherwise, the tensions will continue to prevent India from pursuing a more active regional and interregional role in Central-South Asia. India and China should both be interested in normalization of their relations and stability in the broader region—a major imperative for their unimpeded emergence as great powers.

Finally, India is stuck in the mind frame of its nonalignment tradition, which is potentially detrimental to its projected standing as the largest economy and a major military power in the decades ahead. While it is rightly avoiding a strategic partnership bordering on exclusivity and an antagonistic relationship toward outsiders, it is missing opportunities to join forces or align policies with actor(s) that could advance its influence where it is severely lacking. India still has that opportunity and should consider the benefits and challenges that such a partnership could generate for its global and regional positions. If it views Washington as a potential partner, it needs to address a domestic political discourse viewing the West as the colonizer, which is associated with its nonalignment tradition and struggle against colonialism.[103] India would need to address related constraints inherent in its strategic culture and foreign policy, in the same way Washington would need to imbue with more substance its cooperation with Delhi. After all, only in the aftermath of India's 1998 nuclear test did the U.S. engagement with India start to assume a serious diplomatic outreach rather than predominantly a military collaboration.[104]

Developing a strategically closer partnership with the United States could be key for India's future global and regional role as a great power. As India, the United States faces a number of challenges projecting its influence globally and in Central Asia. But it also has a number of attributes making it a desirable partner for India on the world stage and in the region. Both are democracies and dynamic powers, concerned about China's rise, Russia's resurgence, and stability of Central-South Asian countries facing external pressures and domestic challenges ranging from the lack of economic development to the lack of democratic experience. Both seek a peaceful power transition from the "West to the rest." Washington increasingly views Delhi as a trustworthy actor that could help it retain its leading position in the world. India, in turn, increasingly considers the United States as a platform for its advancement as a great power.

More and more, India relies on the United States to provide it with latest defense technology and organizational models to develop its arms industry and economic processes.[105] In 2005, Delhi and Washington concluded a 10-year-long defense partnership deal allowing for sale of U.S. ballistic missile defense systems and fourth-generation fighter aircraft to India.[106] More recently, they signed an additional nuclear agreement, despite India not being a member of the NPT. Both are now increasingly viewing their cooperation as a key pillar of global and regional stability, conducting joint military exercises and expanding their economic and security cooperation. Both are pursuing similar yet constrained agendas of integrating Afghanistan into Central-South Asia and reconnecting the two regions by supporting geopolitically monumental projects, such as TAPI and CASA-1000, as part of the NSRS and the "connect" policy. Along with Kabul, they further agreed to hold trilateral consultations on promoting Afghanistan's development.[107] But the reality is that neither the NSRS nor the "connect" policy is capable of achieving set objectives faster and more effectively than a closer alignment of policies and engagement between Delhi and Washington designed to generate policy synergies in the wider region.

IV: UNLOCKING AND LEVERAGING INDIA'S REGIONAL POTENTIAL

Regardless of whether India and the United States enter a strategic partnership agreement focused explicitly on Central Asia, they can start, separately or in concert, executing the following objectives to unlock India's potential and enabling Washington to enhance its own: mitigating Af-Pak security challenges; shaping Iran's geopolitical role; fostering Sino-Indian cooperation; and exploiting grand strategies and regional connectivity initiatives of other actors.

Mitigating Af-Pak Security Challenges.

Support India's Reconstruction Effort and Military Involvement in Afghanistan.

Both Washington and Delhi focus on Afghanistan as the node of interregional reconnection, reinforcing the imperative of putting Afghanistan on the track of stable development. China's growing presence, India's lack of resources but a strong intent to play a bigger role, and the U.S. withdrawal but a strong desire to shape Afghanistan's future make it a major task. Washington and Delhi should develop and pursue joint economic, political, and military operations and projects to support Afghanistan's reconstruction and external integration. The parties should upgrade the trilateral consultations (Kabul, Delhi, Washington) by advancing joint and coordinated rather than individual consultations-based action in both economic and military spheres.

Shape Sustainable Political and Military Base in Kabul Favorable to Delhi.

Washington should use freed-up resources after the withdrawal from Afghanistan to shape a sustainable political and military base in Kabul favorable to Delhi. The military's contribution to this effort is crucial, given its pronounced role in the U.S. overall strategy in Afghanistan. The U.S. military should leave behind a military presence and influence sufficient to pursue redefined U.S. objectives in the region; support Afghan security and military capacity; enhance Delhi's military presence in Afghanistan; and advance U.S.-Indian military collaboration in Central-South Asia to complement their growing maritime military cooperation. The U.S. military should leverage its military engagement with India in Afghanistan as a platform to develop India's military intervention capabilities and advance interoperability between U.S. and Indian forces. After all, the U.S. military may support, directly or indirectly, India's military operations or pursue its own military missions in both Afghanistan and the region in the future. In the process, Washington should mitigate Delhi's perceptions of U.S. bias toward Pakistan and encourage deeper U.S.-Indian military cooperation in Afghanistan with a view to expand it to military and other domains in their relations with other states in Central-South Asia.

Enhance Military-to-Military Counterterrorism and Counternarcotics Cooperation.

The reconnection of South Asia with Central Asia increasingly requires transnational cooperation in fighting terrorism and narcotrafficking. This task is critical, considering the role of Afghanistan and Pakistan as hotbeds of homegrown and transnational narcotrafficking and terrorist networks and the potential of the Islamic State (IS) to build its capabilities in Central-South Asia. The U.S. military and intelligence services should pursue enhanced collaboration with Indian, Pakistani, and Central Asian counterparts to tackle both threats. In case of improved ties with Iran, U.S. and Indian militaries should tap into Iran's resources in the fight against IS in Afghanistan and Central-South-Asia. In the process, the U.S. military should carve out a stand-alone platform of military cooperation between the United States and India focused specifically on Central Asia, not just Afghanistan. Working through a prospective partner that has only limited strategic potential in Central Asia, such as India, may allow the U.S military to advance its goals in Central Asia with less resistance from Russia. Moscow has long perceived U.S. efforts to advance counternarcotics and counterterrorist cooperation in the region as a ploy by Washington to promote a stronger military presence in Russia's backyard.

Encourage Af-Pak-Indo Trilateral Economic Cooperation.

Indo-Pakistani tensions remain a major obstacle in reintegrating Central and South Asia, as well as in stimulating intraregional integration in South Asia. In addition to supporting TAPI and CASA-1000, the United States and India should advance trilateral projects in the energy, trade, and transit spheres. Unlocking trade between Pakistan and India would contribute to reconstruction and development of countries in South Asia, while boosting trade from and to Central Asia and enhancing interregional connectivity. Washington and Delhi should actively lobby Kabul and Islamabad to facilitate the implementation and expansion of the Pakistan-Afghanistan Trade and Transit Agreement, specifically as it concerns the enabling of access for India to export goods to Afghanistan via Pakistani territory and the potential involvement of India in the deal in a more expanded, region-wide framework. The role of the U.S. military in the process should rest on the task of advancing confidence-building measures and common understanding of security challenges and opportunities, with a view to support civilian efforts. It should rely on military exercises and educational exchanges in both bilateral and, ideally, trilateral frameworks of military-to-military engagement to promote related objectives.

Facilitate Resolution of Conflict in Kashmir.

The conflict in Kashmir plagues Indo-Pakistani and Sino-Indian relations, leaving room for escalation that could involve parties beyond the three nuclear-armed powers. It also continues to damage the prospects of national and regional development when pressures for connectivity are only bound to rise, as India and China rapidly expand their economic reach. Washington and Delhi should put the issue of resolving the Kashmir conflict on a high agenda in order to reduce regional tensions and increase prospects of regional collaboration. As part of this effort, the U.S. military should develop a better understanding of regional security and economic challenges and their implications for military strategies of concerned powers. It should also develop contingency and special forces operations scenarios in case of conflict escalation, while undertaking initiatives aimed at promoting confidence-building measures, deescalating hostile rhetoric, decreasing militarization, and advancing border management cooperation involving India, Pakistan, and China. It should further work with civilian agencies within the U.S. Government to promote a mutually beneficial border regime in the region as a short-term arrangement, allowing for transcontinental commerce to flourish once again. Looking long-term, the United States and India should work with China, Pakistan, and the people of Kashmir to facilitate a long-term, durable arrangement in the region. In a situation so complex, Kashmir's independence could be a logical, simple, and eventually acceptable solution to the puzzle that has long kept the three nuclear-armed powers on their toes.

Advance India's Geopolitical and Geo-economic Profile in Tajikistan.

India's growing involvement in Afghanistan is a welcome development, but Delhi should also cultivate another node of connectivity, this time in Central Asia proper. While Afghanistan represents such a connectivity node on the northwestern tip of South Asia threatened by instability, Tajikistan represents such a connectivity hub on the southeastern tip of Central Asia threatened by separatism, regionalism, and cross-border militancy. Bordering Afghanistan, Pakistan, and China and dependent on security assistance from Russia and economic aid from China, Tajikistan offers India a platform to reconnect Central and South Asia along the south-north axis, but only if Tajikistan remains whole and integrated from within and without. Washington should foster quadrilateral economic and military cooperation involving Afghanistan, Tajikistan, India, and the United States. As part of this framework, the U.S. military should strive for enhanced military collaboration with Dushanbe and encourage the Tajikistani military to pursue expanded military cooperation with India. The U.S. and Indian military engagement with Tajikistan should advance Tajikistani military capacity to defend the country's borders, given the threat of cross-border militancy from Afghanistan. They should also promote more effective counternarcotics and counterterrorism programs specific to Tajikistan, including as part of military exercises in bilateral and trilateral settings.

Shaping Iran's Geopolitical Role.

Making Iran a Regional Balancer: Welcoming India's Role in Iran and Central Asia.

The 2014-15 P5+1 (China, France, Russia, the United Kingdom, the United States, and Germany) nuclear talks and new geopolitical and geo-economic dynamics and imperatives over recent years have facilitated prospects for a détente between Iran and the United States, which may well evolve into a strategic relationship between Tehran and Washington, potentially within the next 2 decades. Shaping Iran's geopolitical role, set to expand considerably in case of the détente, is critical for Washington and Delhi, given the dynamism displayed by other powers in Eurasia, such as Turkey, Russia, and especially China. Washington and India should ensure that they are a part of Iran's evolution, not its demise. With that in mind, they should start thinking about how to shape Iran's role as a future regional balancer jointly, which would welcome India's expanding role in Iran and Central

Asia and help maintain a balance of power in the Middle East. Specifically, they should identify ways in which Tehran's geopolitical role can be helpful in advancing their interests in Central Asia, including as they relate to other powers and the fight against transnational terrorism and narcotrafficking. In the military sphere, and in the conditions of a U.S.-Iranian détente, the U.S. military should encourage the expansion of Indo-Iranian cooperation in Afghanistan, Central Asia, and the Persian Gulf. The U.S. military should stand ready to calibrate its own infrastructure in Eurasia and leverage Iran's military capabilities to pursue new logistical and operational objectives in a transformed geopolitical environment of the Greater Middle East. By adjusting its military infrastructure requirements, the U.S. military can dedicate its resources to new missions and goals as they relate to China's and Russia's military infrastructure advances in Central-South Asia, the Caspian, the Persian Gulf, and the Indian and Pacific oceans.

Advance U.S.-Iranian and Indo-Iranian Energy, Trade, and Transit Cooperation.

As does India, the United States faces an assertive China, which is invested heavily in energy-rich Iran and seeks to bring Tehran into its fold as part of its "belt" initiative, especially once all sanctions are removed and China finds itself on a shopping spree. Washington and Delhi are either not present in Iran or are far behind China. Meanwhile, the United States, India, and Iran are all somewhat isolated from Central and South Asia, despite immense opportunities for trilateral engagement. Where possible, the parties should identify potential opportunities of cooperation early on to obtain and retain whatever advantage they can in case of a U.S.-Iranian nuclear deal and détente. In the process, Washington and Delhi should act on Iran's concerns about China's dominance in the country's energy sector and advance their presence in Iran's economy to facilitate their policies in the Middle East and Central Asia, where Iran will expand its presence in case of a U.S.-Iranian détente. As India and Iran start to build new and expand existing links with regional countries, cooperating with each other, including as part of IPI, will prove beneficial. But whatever economic links they pursue, the parties should ensure that proper security arrangements are in place to facilitate the development and expansion of such connections. The role of the U.S. military in developing combined military contingencies, advancing common understanding with Indian and Iranian militaries, and deterring state and nonstate actors from targeting this infrastructure would be crucial, especially during crises and armed conflicts involving regional partners and rivals.

Encourage Indo-Iranian Cooperation in Reconstruction of Afghanistan.

Iran is not only a key to integration of India with Central Asia. It is also a crucial link in Delhi's reconnection with the Greater Middle East. Bordering Afghanistan in the east, sharing a cultural heritage with Afghanistan, and given the history of its support to U.S. objectives in Kabul and interest in expanding its reach in case of improved ties with the West, Iran could represent a major asset to the United States and India. Its role, channeled in circumstances of growing strategic collaboration between Tehran and Washington, could promote U.S. and Indian interests and policies in Afghanistan and Central-South Asia. Neither party benefits from destabilization in Afghanistan, where they pursue reconstruction efforts. India and Iran already seek to expand energy and trade ties with each other via Afghanistan as part of their economic policies toward Central Asia. Washington should support the Indo-Iranian collaboration aimed at stabilization and integration of Afghanistan into Eurasia by advancing trilateral economic and counterterrorism collaboration focused on Afghanistan, Central Asia, and the Middle East. In the process, and in the conditions of a U.S.-Iranian detente, the U.S. military should share intelligence and work with its Iranian and Indian military counterparts in an effort to degrade terrorist and militant networks operating in and out of the Af-Pak region.

Fostering Sino-Indian Cooperation.

Advance Resolution of Border Disputes and Demilitarization in Border Areas.

Border disputes strain Sino-Indian ties, impeding economic linkages not only between the two powers but also between China and South Asia, on the one hand, and between India and Central Asia, on the other. The United States and India should dedicate more resources to advancing the resolution of Sino-Indian border disputes and demilitarization along the border areas, but without portraying or positioning Washington as an arbiter. The U.S. military's role here comes down to the following major tasks: advancing demilitarization in Sino-Indian border areas; promoting common understanding between Chinese and Indian militaries as part of bilateral and trilateral engagement; holding bilateral or multilateral military exercises with Central Asian, Chinese, Indian, and Pakistani armies; promoting educational exchanges; sharing intelligence and security assessments; and planning joint and combined contingency operations in case of conflict escalation in disputed border areas. Resolving these border disputes would contribute to normalization of relations between two major Asian rivals, helping the United States uphold systemic stability. It would also enhance China's and India's roles as interregional economic engines, contributing to development and integration within Eurasia. The parties could then pursue and expand economic links via Xingjian and Kashmir more effectively.

Facilitate Joint and Combined Sino-Indian Resource Acquisition Strategies.

China's "go global" campaign has achieved a more active and far-reaching resource acquisition strategy globally, including in Central Asia. As a country trailing behind, India needs China's cooperation in select parts of the world to advance its own resource acquisition strategy. This especially concerns Central Asia, a region of growing importance to the resource-hungry Asian giants where Beijing outbids Delhi. Both already have experience pursuing joint bids to acquire stakes in oil and gas fields in various parts of the world. But this experiment is limited, as both countries continue to perceive each other as long-term strategic rivals.

The United States should encourage Delhi and Beijing to participate in combined resource development, production, and delivery projects in Central Asia, both on a diplomatic and financial level. A result is not only an enhanced ability for India to access Central Asian markets, but also an improved and more trustworthy relationship between India and China as a source of global stability. At the same time, the U.S. military should support and encourage Delhi's incipient efforts to build a regional, and eventually global, military infrastructure to protect its expanding economic interests and presence around the world. As it does so, it should focus on identifying common threats and opportunities for military engagement involving all three parties so as to lessen the prospect of armed conflict amid the global race for resources and on devising joint political and military approaches to address crises and conflicts induced by resource scarcity, environmental degradation, and natural disasters.

Stimulate Confidence-Building Measures and Shared Regional Security Approaches.

The rise of China and India is a source of major challenges and opportunities for the global and regional security system. As the two countries' profiles expand in Central and South Asia, both China and India should develop confidence-building measures and shared regional security management approaches. The alternative is a lingering suspicion and potentially explosive conflict between the two countries, which are only starting to learn the ropes of projecting influence globally in the conditions of the 21st century. The United States should encourage the two powers to develop common platforms of cooperation in Central-South Asia but seek to participate in such arrangements where practical to prevent an unlikely but potential emergence of an unfriendly "Chindia." Washington should support India's membership in the

SCO, allowing India to advance its regional presence, check China's rise, and boost interregional reconnection. A security framework of cooperation promoted by and involving India and China would allow for a sustainable evolution of different parts of the rapidly developing and developed Asia. In the process, the U.S. military should focus on promoting the idea of and helping design shared security and military arrangements and responses to regional and/or domestic crises in Central Asia in an institutional setting. The U.S. military should consider ways of cooperating with the SCO in areas such as counterterrorism, counternarcotics, border management, and the fight against cross-border militancy. Provided India becomes a member of the SCO, the U.S. military should work with Indian military to advance related objectives, while promoting a shared approach to security management and discoursing competitive dynamics that are prone to conflict.

Exploiting Grand Strategies and Regional Connectivity Initiatives.

Advance India's Role in Grand Strategies of Central Asian Republics.

Central Asian states view the role of outside powers as crucial in their grand strategies. Currently, India does not feature prominently in these strategies, but Washington and Delhi can address this deficiency by advancing the role of India as a future great power. In the process, they should emphasize to the regional states the importance of developing the southern vector of connectivity by advancing strategic ties with India, while pointing to a possible U.S.-Indian partnership in facilitating this dynamic. As part of the U.S. strategy, the U.S. military should promote the geopolitical role of India and the United States in the military doctrines of the regional states as a component of their grand strategies; leverage India's unimposing regional posture and counterterrorism capabilities to advance military ties with the republics; develop security arrangements to safeguard infrastructure of transregional projects with participation of Central Asian states, Afghanistan and, under the right conditions, Iran; identify and help address security vulnerabilities of individual republics as part of bilateral and multilateral military collaboration, including through intelligence sharing and arms sales/transfers; increase the frequency of counterterrorism and disaster relief military exercises; and increase the quota of students and scholarships for Central Asian students in U.S. and Indian military academies. Pursued together, these efforts would reinforce the message that the regional states have options to pursue strategic engagement with distant actors, as well.

Align and Combine Parts of U.S. NSRS with India's "Connect" Policy.

Both the NSRS and the "connect" policy share similar objectives, but neither one is solely capable of achieving its objectives faster, more efficiently, and in a more sustainable manner. Aligning, combining, and jointly pursuing components of the two initiatives would generate policy synergies and achieve outcomes faster and more effectively. India and the United States could combine the U.S. "software" with Indian "hardware" or vice-versa dependent on a project. U.S. institutional and trade facilitation expertise in combination with infrastructure development and investments by India would help complete a Central-South Asian economic bridge faster. Besides other benefits, doing so would enhance the symbolism of India and the United States working together and demonstrate the commitment of both powers to regional economic development and security after 2016 when Washington and its allies plan to withdraw from Afghanistan. But closer U.S. economic engagement with its prospective partners would require a new role for the U.S., Indian, Afghan, and Central Asian militaries in safeguarding their interests. As part of its military-to-military engagement with its regional counterparts, the U.S. military should facilitate security conditions conducive for effective and uninterrupted functioning of cross-regional projects and initiatives, such as TAPI, CASA-1000, and other existing and potential initiatives being advanced by the parties in the broader region. It should develop security arrangements and contingencies for engaging individual, joint, or combined military and security forces to safeguard these assets, especially during regional crises or armed conflicts threatening this infrastructure.

Channel China's "Belt" and Other Connectivity Initiatives in South-West Direction.

China's "belt" initiative is primarily an east-west corridor via Central Asia. And while the CPEC facilitates a south-north connection, it does not involve Central Asian states or India. Washington and Delhi should advance the participation of Afghanistan, India, the United States, and Central Asian republics in China's "belt," potential expansion of the CPEC, and other connectivity projects to foster east-southwest and south-northwest linkages. Washington and Delhi should focus on positioning Kashmir and Xingjian as energy, trade, and transit connectors and shape their evolution in conjunction with tasks seeking the resolution of border disputes and relaxation of tensions in the region. Such an approach would create openings for the United States, India, Afghanistan, and Central Asian states to develop north-south and south-north linkages in addition to east-west and west-east connections and do so in collaboration with China and Pakistan. As such U.S.-Indian engagement in transregional projects grows, and given the conflict potential of the areas where these projects operate, the U.S. and Indian militaries should synch their views and practices to advance security and stability in the region. They should also work with their counterparts in China and Pakistan to mitigate any suspicions associated with the growing U.S.-Indian economic and military engagement in the region and, ideally, advance common understanding and shared security arrangements. Focusing on counterterrorism as a first step could be the best platform for deepening military-to-military engagement, as all four actors share a history of struggle against terrorism.

Facilitate the INSTC Linking Russia and India via Central Asia.

As Russia pulls Central Asia north and China pulls it east, it is imperative that India and the United States ensure that their connectivity strategies open up new vectors of economic development, military engagement, and strategic collaboration for Central Asian republics without imposing them or denying the republics a choice of pursuing linkages in other directions.

Washington and Delhi should facilitate and expand the INSTC, necessarily linking the EEU, Russia, and India via the Caspian region and Central Asia. This would help develop a vertical line in Eurasia strong enough to approximate, if not match, the horizontal line of rapidly expanding east-west economic linkages. Otherwise, the balance of forces in Eurasia will rest on a skewed axis, constraining strategic options for regional economies and external players alike. To assist the overall U.S. goals of advancing transcontinental connectivity, the U.S. military should work with its partners in NATO, Russia, India, and Central Asia to advance security of the areas where INSTC would operate in its existing and expanded format. In the northern tier of the INSTC, the U.S. military effort should concentrate on the task, seeking to mitigate interstate tensions in the region stemming from the unresolved status of the Caspian. In the southern tier, it should work with its regional counterparts to reduce security threats, such as cross-border militancy, transnational terrorism, and narcotics and human trafficking, which threaten the economic flows generated by the INSTC.

V: INDIA'S LONG-TERM PRESENCE IN CENTRAL ASIA: FROM ASPIRATIONS TO INFLUENCE

From the 16th to the 18th century, India and Central Asia interacted as part of an "interimperial system" that involved the Ottoman, the Mughal, the Safavid, and the Shaibanid empires. A common "Persianate code of conduct" going beyond religious affiliations defined interimperial communication and diplomacy, linking the empires. But the Great Game, the Russian revolution, the creation of the Soviet Union, the fall of the Ottoman Empire, and the de-Persianization of India have made the borders between empires less permeable.[108] Even the end of the Cold War and the emergence of independent Central Asian states made this interaction difficult, as the newly independent countries cautiously embarked on their quest for a place on the political map.

But a quarter of a century afterward, Central Asia is opening up, albeit still slowly, assuming a growing strategic importance for neighbors and the rest of the world. India, which had more or less interacted with the region until the creation of the Soviet Union, has become increasingly engaged in Central Asia over the years. But unlike Russia, which has been projecting its influence in and out of the region for a century and a half, India is a latecomer to the region's 21st century "Great Game." And unlike China, it is starting its great power ascendance from a lower base and at a slower pace, explaining its policy failures globally and in Central-South Asia.

Select elements of India's strategic culture and geopolitical constraints account for its delayed arrival and limited presence in Central Asia compared to great and middle-ranked powers alike. At the same time, a number of attributes make it a major partner for the regional states in the long run, capable of standing up to a potential regional hegemon—China. While Delhi may choose to rely on its own resources to advance its strategic position in Central Asia on par with Russia, China, or the United States, it is far better off partnering with Washington on a strategic level, given constraints of its strategic culture and geopolitical challenges it confronts. Without collaboration with the United States, its strategic positioning in Central Asia and the world will be slower, while its efforts to turn its regional aspirations into influence will be severely constrained.

Ever since it officially shed the principle of nonalignment in its foreign policy after the collapse of the Soviet Union, India has failed to define its role for the region or the world as an aspiring great power or serve as a model of development. As a result, it has no clear direction to pursue and no appealing model to offer to its neighbors, including in Central Asia where the regional republics look more to Russia and China for guidance and assistance on economic, security, and development matters. Nor has it yet fully embraced the concept of national interests as the guiding principle of its policy, even if one could justify the importance of nonalignment in advancing its perceived national interests during and after the Cold War. Unlike the United States or China, India has no allies. But its efforts at cultivating such a relation with Vietnam[109] is a promising sign that Delhi should replicate in Central Asia, where Tajikistan and Kazakhstan could serve such roles on the southernmost and northernmost points of Central Asia.

India's lack of success in advancing its presence in Central Asia is not the result of its own failures alone. Its "Connect Central Asia" policy confronts a disconnected broader region of Central and South Asia, whose subregions lack connectivity within and between each other and witness ethnic tensions and rivalries for primacy among constituent states. Unlike Russia or China, India does not border Central Asia, having to rely on airlifts from Delhi and Dubai, volatile Afghanistan, and relatively isolated Iran to trade with the region. It also faces an unstable Afghanistan and unfriendly Pakistan, whose evolution is key to its regional ambitions. Meanwhile, its tensions with Islamabad and Beijing further limit its reach in Central Asia. As a result, Delhi is confined to South Asia, unable to unleash its potential as an aspiring great power.

As India's rival for global primacy and resources, the rising China presents a formidable strategic challenge to India. This challenge is especially pronounced in India's immediate neighborhoods of Central and South Asia, where Beijing has been making increasingly successful, far-reaching strides. China's head start in economic reforms and its already strongest position in Central Asia have contributed to its relatively stronger geopolitical and geo-economic position in Central Asia, reinforcing Delhi's perception of Beijing's intent to encircle India, tie it down in South Asia, and prevent it from enhancing its presence in Central Asia. This has prompted India to counter China's moves in an attempt to retain its position, including in Central and South Asia, where it has been bolstering its partnerships with regional countries. But the challenge presented by China is not just that of rivalry for resources and influence. It is also one of finding political will to cultivate a collaborative competition between the two rising powers. India should thus partner with Beijing where possible to advance its regional position.

India's failures attributed to its strategic culture, geopolitical constraints, and rivalries of local and external powers within Central and South Asia require Delhi to enhance its effort to connect with the region—something it has yet to achieve as part of its "connect" policy. Launched in 2012, the policy has sought to link Central and South Asia by focusing on the development and integration of Afghanistan from within and without, thereby opening new doors for India to import strategically vital resources for its resource-hungry economy and to project its rapidly growing power not only out of South Asia but also Central Asia. India's "connect" policy is a reflection of both geopolitical constraints it confronts and geopolitical solutions it employs to overcome these constraints and extend its reach. Despite its proactive stance and focus on action rather than observation, the "connect" policy has a long way to go before it delivers. In large part, this has to do with instability in Afghanistan and Pakistan—the countries on which India relies to fully connect with Central Asia and that teeter on the brink of potential disintegration.

India's "connect" policy also has to contend with relatively less constrained policies of other actors in Central Asia, big and small. The attained presence of these actors, especially Russia and China, but also Turkey and individual EEU members, make Delhi's desired regional goals look more like aspirations and ambitions rather than attainable gains and real prospects. To advance its presence in Central Asia, India can and should capitalize on favorable views of the regional republics and areas of potential engagement that exist between them, especially in energy, weapons procurement and defense, transit infrastructure development, and trade facilitation. It should also engage rather than shun its perceived rivals. Doing so would advance development in and between Central and South Asia, helping it unlock its strategic potential and enabling the regional republics to develop a southern vector of development and build their external balancing capacity. While a latecomer in the region, India is not unwelcomed by Central Asian states that are eager to diversify their ties and secure access to the Indian subcontinent and Indian Ocean.

But this approach is not enough, considering India's lost ground since the collapse of the Soviet Union. A search for allies and partners is a strategic imperative for India—a task bordering on a feat, as India is yet to discard the legacy and policy inertia of its nonalignment ideology and embrace the national interest as the guiding principle of an active rather than passive foreign policy. While Delhi should strive for cooperation with its current and long-term rivals and challengers such as Pakistan and China, it should develop a weighty strategic partnership with the United States. As the world's strongest power with a democratic system and limited role in Central Asia, Washington is a natural partner for Delhi in a region dominated by authoritarian Russia and China. India's lack of imperial history, democratic tradition, history of nonalignment, and shared concern about China's emergence make Delhi a prospective partner for the Unites States globally and regionally. Joining forces in Central Asia would allow both to advance and sustain their long-term regional positions vis-à-vis other powers. This is critical because both enjoy limited presence in Central-South Asia, pursue similar agendas, and confront similar global and regional threats posed by state and nonstate actors. As they consider such a prospect, they should unlock and leverage India's regional potential by pursuing the following objectives, which should also uncover the U.S. strategic potential in Central-South Asia.

First, the United States and India should work together to mitigate security challenges in the Af-Pak region by focusing on development, reconstruction, and integration of Afghanistan into the broader region and by encouraging Pakistan to clamp down on homegrown militant and terrorist groups. This objective rests on the execution of a number of following tasks, allowing India to connect with Central Asia more safely, efficiently, and in a sustainable manner: supporting and expanding India's reconstruction and involvement effort in Afghanistan; encouraging Af-Pak-Indo economic cooperation and relaxation of political tensions; facilitating resolution of the conflict in Kashmir; and advancing India's profile in Tajikistan, which borders Pakistan and China and could serve as Central-Asia's receiving end of interregional reconnection.

Second, the United States and India should shape Iran's geopolitical role in the world and Central-South Asia provided Washington and Tehran conclude a nuclear deal and work toward a détente. In the process, they should encourage Iran to play a role of regional balancer vis-à-vis Turkey in the Middle East and China and Russia in Central Asia, which would welcome India's growing presence in the Iranian economy dominated by China and facilitate India's strategic presence in Eurasia. They should further advance Indo-Iranian energy, trade, and transit cooperation with a focus on involvement of Central Asian states and Afghanistan, in part by implementing the long-overdue IPI project and encouraging Iran's participation in resource deliveries from Turkmenistan and the Caspian to Pakistan and India. Moreover, they should encourage Indo-Iranian cooperation in rebuilding and integrating Afghanistan from within and with Central-South Asia, especially given the potential of the country's disintegration after 2016.

Third, while Washington and Delhi pursue co-engagement with China, they should not lose sight of the task to foster Sino-Indian cooperation to ensure a smooth reconnection of Central and South Asia and a peaceful global environment conducive for unlocking India's long-term potential in the world and the broader region. Harnessing and channeling China's rapid push into Eurasia is critical for interests of both the United States and India, given Washington's declining global profile and Delhi's growing but constrained global role. To that purpose, the United States and India should advance the resolution of Sino-Indian border disputes; facilitate the pursuit of joint and shared Sino-Indian resource acquisition strategies, especially in the immediate neighborhoods of Central and South Asia; and stimulate confidence-building measures and shared security approaches, including by supporting India's membership in the Beijing-led SCO.

Finally, the United States and India should not only imbue their connectivity initiatives with much-needed commitment, funding, and leadership, but they should also exploit grand strategies and connectivity initiatives of other powers to facilitate Eurasian internal and external integration with the global economy, while opening up opportunities for India's south-north economic push. To achieve this, Washington and Delhi should advance India's role in grand strategies of Central Asian republics by highlighting concerns centered on the rise of China and the benefits of a strategic U.S.-Indian partnership for the region. They should also align parts of their connectivity initiatives to achieve faster and more effective regional outcomes. They would further benefit from channeling China's "belt" and other connectivity initiatives in southward and south-northeast directions. Finally, the parties would do well to facilitate north-south trade and transit linkages, including as part of an expanded INSTC already linking Russia, the EEU, and India, that should involve Central Asian countries, as well as Afghanistan and Pakistan.

Whether India will rely on its own resources—to be expanded with the rise of its economy—or seek a partnership with the United States, or both, remains to be seen. What is clear is that India's geopolitical role is set to expand considerably in the years and decades ahead. Such expansion will generate concerns for neighbors and distant actors alike, creating conditions for instability despite any benefits of cooperation between them and India as one of the largest economies. Washington and Delhi should ensure they stay engaged in Central Asia and enhance their positions amid a power struggle unfolding between outside powers in this increasingly critical part of the world. To do this, they should cultivate a strategic partnership that makes Central Asia its major pillar. Until then, neither Delhi nor Washington is likely to succeed.

The role of the U.S. military in the process, either as part of the U.S.-Indian strategic partnership or as part of U.S. overall efforts to cultivate such a partnership with Delhi, will be critical. Rightly so, because a lot is at stake for the United States in this seemingly remote backwater. As an area of growing external rivalries, the region of Central-South Asia is a source of both traditional and nontraditional security threats to U.S. national interests, be that in the political, military, or even economic domain. From interstate conflicts to transnational terrorism and from Russia's attempts to reestablish geopolitical control to China's efforts to achieve economic dominance, the region is a focal point of intersecting challenges and opportunities that the U.S. military should be better positioned to address and leverage in support of U.S. interests. Pursuing those objectives as part of U.S. economic, political, and military efforts would assist India in unlocking its regional strategic potential and help Washington unlock its own.

ENDNOTES

1. Aaron L. Friedberg, *A Contest for Supremacy: China, America, and the Struggle for Mastery in Asia*, New York: W. W. Norton & Company, 2011, p. 25.

2. *Ibid.*, p. 26.

3. Sudeepto Adhikari, Akbouri Radha Krishna Sinha, and Mukul Kamle, "India's Changing Geopolitical Code: an Attempt at Analysis," *Geopolitics Quarterly*, Vol. 3, No. 4, Winter 2008, p. 26.

4. *Ibid.*, p. 19. Also see Harsh V. Pant, *The Growing Complexity of Sino-Indian Ties*, Carlisle, PA: Strategic Studies Institute, U.S. Army War College Press, 2014, p. 3.

5. Emilian Kavalski, *India and Central Asia: The Mythmaking and International Relations of a Rising Power*, New York: I. B. Tauris Publishers, 2010, p. 30.

6. Sebastien Peyrouse, "Domestic and International Articulations of the Indian Involvement in Central Asia," in Marlene Laruelle and Sebastien Peyrouse, eds., *Mapping Central*

Asia: Indian Perceptions and Strategies, Burlington, VT: Ashgate, 2011, pp. 75-79.

7. Pratap Bhanu Mehta, quoted in Emilian Kavalski, *Central Asia and the Rise of Normative Powers: Contextualizing the Security Governance of the European Union, China, and India*, New York: Bloomsbury, 2012, p. 127.

8. A. Vanaik, "Making India Strong: The BJP-led Government's Foreign Policy Perspectives," in J. McGuire and I. Copland eds., *Hindu Nationalism and Governance*, New Delhi, India: Oxford University Press, pp. 390-391.

9. Quoted in *Ibid.;* Kavalski, p. 128.

10. S. Mahalingam, "India-Central Asia Energy Cooperation," in K. Santhanam and R. Dwivedi, eds., *India in Central Asia: Advancing the Common Interest*, New Delhi, India: Anamaya Publishers, 2005, p. 133. Also see Kavalski, p. 139.

11. Kavalski, pp. 140-141.

12. Vishakha N. Desai, "From Statecraft to Soulcraft," in Clay Chandler and Adil Zainulbhai, eds., *Reimagining India: Unlocking the Potential of Asia's Next Superpower*, New York: Simon & Schuster, 2013, pp. 302-303.

13. Peyrouse, "Domestic and International Articulations of the Indian Involvement in Central Asia," pp. 79-80.

14. For more discussion, see Kavalski, pp. 45, 67-68.

15. Ajay Patnaik, "Framing Indo-Central Asian Relations, 1990s-2000s," in Laruelle and Peyrouse, p. 98.

16. *Ibid.*, pp. 98-99.

17. Marlene Laruelle, "Foreign Policy and Myth-making: Great Game, Heartland, and Silk Roads," in Laruelle and Peyrouse, p. 16. Also see Sanjay Kumar Pandey, "India and the Importance of Central Asia," in P. L. Dash, ed., *India and Central Asia: Two Decades of Transition*, New Delhi, India: Oxford University Press, 2012, p. 2.

18. Nimmi Kurian, *India-China Borderlands: Conversations Beyond the Center*, New York: Sage, 2014, pp. 66-83, 91.

19. George Friedman, "The Geopolitics of India: A Shifting, Self-Contained World," Stratfor, December 16, 2008.

20. Robert D. Kaplan, *The Revenge of Geography: What the Map Tells about Coming Conflicts and the Battle against Fate*, New York: Random House, 2012, p. 254.

21. *Ibid.*, p. 233.

22. Meena Singh Roy, "India's Policy Toward Central Asia: The Pakistan Factor," in Laruelle and Peyrouse, p. 175.

23. Ajay Darshan Behera, "The Resurgence of the Taliban and Regional Security Concerns," in Laruelle and Peyrouse, p. 203.

24. Roy, p. 175.

25. "Fatalities in Terrorist Violence in Pakistan 2003-2015," The South Asia Terrorism Portal, accessed June 18, 2015.

26. Robert D. Kaplan, *Monsoon: The Indian Ocean and the Future of American Power*, New York: Random House, 2010, p. 112.

27. Gurcharan Das, the CEO and author, quoted in Fareed Zakaria, "The Rediscovery of India," in Chandler and Zainulbhai, p. 5.

28. Also see Sebastien Peyrouse, "Comparing the Economic Involvement of China and India in post-Soviet Central Asia," in Marlene Laruelle, Jean-Francois Huchet, Sebastien Peyrouse, and Bayram Balci, eds., *China and India in Central Asia: A New "Great Game?"* New York: Palgrave Macmillan, 2010, pp. 155-156.

29. Basudeb Chaudhuri and Nampreet Sethi, "India-China Interactions in Central Asia through the Prism of Paul Kennedy's Analysis of Great Powers," in Laruelle, Huchet, *et al.*, p. 127.

30. Wolfgang Schurer, "A Geopolitical and Geo-economic Overview: On the Rise of China and India as Two Asian

Giants," *The Fletcher Forum of World Affairs*, Vol. 29, Issue 2, Summer 2005, p. 158.

31. For more on China's perceived encirclement of India, see David Scott, "The Great Power 'Great Game' between India and China: 'The Logic of Geography'," *Geopolitics*, Vol. 13, Issue 1-26, 2008, pp. 4-7.

32. For more on the term, see Zhao Gancheng, "The Rise of Chindia and Its Impact on the World System," in Roy Huisken, ed., *Rising China: Power and Reassurance*, Canberra, Australia: The Australian National University Press, 2009, pp. 65-77.

33. Harsh V. Pant, "China's Half-Hearted Engagement and India's Proactive Balancing," in Ian Hall, ed., *The Engagement of India: Strategies and Responses*, Washington, DC: Georgetown University Press, 2014, p. 121.

34. Sanjay Kumar Pandey, "India and the Importance of Central Asia," in Dash, p. 7.

35. Peyrouse, "Comparing the Economic Involvement of China and India in post-Soviet Central Asia," pp. 84-85.

36. Swaran Singh, "China-led Multilateralism in Central Asia: Strategic Implications for India," in Dash, pp. 66-67.

37. Yu Jianhua, "Competition for Caspian Energy Development and Its Relationship with China," in Charles Hawkings and Robert Love, eds., *The New Great Game: Chinese Views in Central Asia*, Fort Leavenworth, KS: Foreign Military Studies Office, 2006, p. 149.

38. For more on India's strategy countering China's perceived policy of encirclement, see Scott, pp. 7-9.

39. Nirmala Joshi, "India-Central Asia Energy Cooperation" in Dash, p. 56.

40. Schurer, pp. 72, 161.

41. Kaplan, *Monsoon*, p. 249.

42. Arun Prakash, "Introduction" in Sam Bateman and Joshua Ho, eds., *Southeast Asia and the Rise of Chinese and Indian Naval Power: Between Rising Naval Powers*, New York: Routledge, 2010, p. 2.

43. Sumit Ganguly and Manjeet S. Pardesi, "Can China and India Rise Peacefully?" *Foreign Policy Research Institute*, Summer 2012, pp. 477-478.

44. Adhikari, *et al.*, pp. 22-25.

45. Letter from Atal Vajpayee to Bill Clinton, as cited in Pant, *The Growing Complexity of Sino-Indian Ties*, pp. 4, 28.

46. *Ibid.*

47. Mushtaq A. Kaw, "Restoring India's Links with Central Asia across Kashmir: Challenges and Opportunities," in Laruelle and Peyrouse, p. 181. Also see Marlene Laruelle and Sebastien Peyrouse, "Managing Regional Geopolitical Uncertainty," in Laruelle and Peyrouse, p. 156.

48. Vinay Rai and William L. Simon, *Think India: The Rise of the World's Next Great Power and What It Means for Every American*, New York: The Penguin Group, 2008, p. 247.

49. Jean-Francois Huchet, "India and China in Central Asia: Mirroring Their Bilateral Relations," in Laruelle, Huchet, *et al.*, pp. 112-113.

50. Roman Muzalevsky, *Central Asia's Shrinking Connectivity Gap: Implications for U.S. Strategy*, Carlisle, PA: Strategic Studies Institute and U.S. Army War College Press, November 2014, p. 37.

51. Ivan Campbell, "India's Roles and Interests in Central Asia," Safeworld Briefing, October 2013, p. 2.

52. Quoted in Zhao Huasheng, "Cooperation or Competition? China and India in Central Asia," in Marlene Laruelle, Jean-Francois Huchet, Sebastien Peyrouse, and Bayram Balci, eds., *China and India in Central Asia: A New "Great Game"?*, New York: Palgrave Macmillan, 2010, pp. 132-133.

53. Peyrouse, "Domestic and International Articulations of the Indian Involvement in Central Asia," p. 86.

54. Raj Kumar Kothari, "India's 'Connect Central Asia Policy': Emerging Economic and Security Dimensions," *Sociology and Anthropology*, Vol. 2, Issue 6, 2014, p. 240.

55. Gulshan Sachdeva, "Indo-Central Asian Economic Relations," in Laruelle and Peyrouse, p. 132.

56. Kothari, p. 242.

57. Peyrouse, "Comparing the Economic Involvement of China and India in post-Soviet Central Asia," pp. 82-83.

58. Nivedita Das Kundu, "India-Russia Strategic Cooperation: Eye on the Future," in Nivedita Das Kundu, ed., *India-Russia Strategic Partnership: Challenges and Prospects*, New Delhi, India: Academic Foundation and Indian Council of World Affairs, 2010, p. 157.

59. S. Varadarajan, "Power Grids and the New Silk Road in Asia," *The Hindu*, July 11, 2005.

60. Sachdeva, p. 137.

61. For more, see Roman Muzalevsky, *From Frozen Ties to Strategic Engagement: U.S.-Iranian Relations in 2030*, Carlisle, PA: Strategic Studies Institute, U.S. Army War College Press, November 2014, p. 38.

62. *Ibid*.

63. Roman Muzalevsky, "India's 'Connect Central Asia' Policy Seeks to Compensate for Lost Time," *Eurasia Daily Monitor*, Vol. 9, Issue 176, September 27, 2012.

64. Marlene Laruelle and Sebastien Peyrouse, "Managing Regional Geopolitical Uncertainty," in Laruelle and Peyrouse, p. 151.

65. For more discussion see Muzalevsky, *From Frozen Ties to Strategic Engagement*.

66. Ahmed Rashid, "Stumbling toward Peace," in Chandler and Zainulbhai, p. 365.

67. Laruelle and Peyrouse, p. 155.

68. Sachdeva, p. 139.

69. Sumit Ganguly and Manjeet S. Pardesi, "Can China and India Rise Peacefully?" Foreign Policy Research Institute, Summer 2012, p. 474.

70. Sreemati Ganguli, "The Revival of the Silk Roads Topic: A Contemporary Analysis," in Marlene Laruelle and Sebastien Peyrouse, eds., *Mapping Central Asia: Indian Perceptions and Strategies*, Burlington, VT: Ashgate, 2011, p. 61.

71. C. Raja Mohan, "Emerging Geopolitical trends and Security in the Association of Southeast Asian Nations, the People's Republic of China, and India (ACI) Region," Asian Development Bank Institute (ADBI) Working Paper Series, No. 412, March 2013, p. 14.

72. Nirmala Joshi, "India-Central Asia Energy Cooperation" in P. L. Dash, ed., *India and Central Asia: Two Decades of Transition*, New Delhi, India: Oxford University Press, 2012, p. 56. Also see Mushtaq A. Kaw, p. 187.

73. Ajay Patnaik, "Framing Indo-Central Asian Relations, 1990s-2000s," in Laruelle and Peyrouse, p. 102.

74. Kaw, p. 187.

75. Harsh V. Pant, "China's Half-Hearted Engagement and India's Proactive Balancing," in Hall, p. 118.

76. P. Stobdan, "Central Asia: India's Northern Exposure," Institute for Defense Studies and Analyses Monograph Series No. 44, New Delhi, India: Institute for Defense Studies and Analyses, May 2015, p. 21.

77. Pant, "China's Half-Hearted Engagement and India's Proactive Balancing," p. 119.

78. For more on mutual perceptions, see Huasheng, pp. 131-138.

79. Jagannath P. Panda, "India's New Look at Central Asia Policy: A Strategic Review," in Laruelle, Huchet, *et al.*, pp. 115-116.

80. Aset Ordabaev, Geopolitika transportnykh koridorov v Tsentralnoi Azii (*The Geopolitics of Transport Corridors in Central Asia*), Astana, Kazakhstan: Institute of World Economy and Politics, the Fund of the First President of Kazakhstan, p. 44.

81. Gulnoza Saidazimova, "Integraziya v Tsentralnoi Azii: realii, vyzovy, vozmojnosti" ("Integration in Central Asia: Realities, Challenges, and Opportunities"), *Tsentralnaia Azia i Kavkaz*, September 2000.

82. Nick Bisley, "Engagement, India, and the Changing International Order," in Hall, p. 190.

83. Marlene Laruelle, "Russia Facing China and India in Central Asia: Cooperation, Competition, and Hesitations," in Laruelle, Huchet, *et al.*, p. 21.

84. Quoted in Kaplan, *Monsoon*, p. 129.

85. Peyrouse, "Domestic and International Articulations of the Indian Involvement in Central Asia," pp. 80-86.

86. I. Iskakov, "Problemy i perspektivy integratsii evraziiskogo prostranstva" ("Issues and Prospects of Eurasian Integration"), ББК 63.3(2)64-6, p. 243.

87. Sebastien Peyrouse, "Comparing the Economic Involvement of China and India in Post-Soviet Central Asia," in Laruelle, Huchet, *et al.*, p. 168.

88. Louise Merrington, "India and China: Strategic Engagements in Central Asia," in Hall, p. 91.

89. Adiljan Umarov and Dmitriy Pashkun, "Tensions in Sino-Central Asian Relations: Implications for Regional Security," Central Asia Series, Watchfield, England: Conflict Studies Research Centre, Defence Academy of the United Kingdom, January 2006, p. 2.

90. D. Kaushik, "India and Central Asia: Past and Present Relations and Future Hopes," *Contemporary Central Asia*, Vol. 1, Issue 1, 1996, pp. 226-239.

91. Sascha Muller-Kraenner, "China's and India's Emerging Energy Foreign Policy," Discussion Paper, German Development Institute, Bonn, Germany, 2008.

92. Peyrouse, "Comparing the Economic Involvement of China and India in Post-Soviet Central Asia," p. 159.

93. *Ibid*.

94. Ordabaev, p. 16.

95. Farkhod Tolipov, "Flexibility or Strategic Confusion? Foreign Policy of Uzbekistan," Uzbekistan Initiative Papers, February 2014, pp. 3-5.

96. Peyrouse, "Domestic and International Articulations of the Indian Involvement in Central Asia," p. 81.

97. Peyrouse, "Comparing the Economic Involvement of China and India in Post-Soviet Central Asia," p. 159.

98. Sanjay Kumar Pandey, "India and the Importance of Central Asia," in Dash, p. 8.

99. Peyrouse, *Monsoon*, p. 197.

100. Stobdan, pp. 48-54.

101. Panda, p. 116.

102. Kavalski, pp. 102-103.

103. Peyrouse, "Domestic and International Articulations of the Indian Involvement in Central Asia," p. 88.

104. Ian Hall, "The Engagement of India," in Hall, pp. 7-9.

105. Cohen, "India and America: Redefining the Partnership," in Chandler and Zainulbhai, p. 337.

106. Hall, pp. 7-9.

107. Kothari, p. 242.

108. Laurent Gayer, "From the Oxus to the Indus: Looking Back at India-Central Asia Connections in the Early Modern Age," in Laruelle, Huchet, *et al.*, p. 127.

109. For more on China's perceived encirclement of India, see Scott, p. 10.

U.S. ARMY WAR COLLEGE

Major General William E. Rapp
Commandant

*****

STRATEGIC STUDIES INSTITUTE
and
U.S. ARMY WAR COLLEGE PRESS

Director
Professor Douglas C. Lovelace, Jr.

Director of Research
Dr. Steven K. Metz

Author
Mr. Roman Muzalevsky

Editor for Production
Dr. James G. Pierce

Publications Assistant
Ms. Rita A. Rummel

*****

Composition
Mrs. Jennifer E. Nevil

Made in the USA
Monee, IL
04 November 2026

67378946R00066